Artificial Intelligence on Dark Matter and Dark Energy

As we prod the cosmos at very large scales, basic tenets of physics seem to crumble under the weight of contradicting evidence. This book helps mitigate the crisis. It resorts to artificial intelligence (AI) for answers and describes the outcome of this quest in terms of an ur-universe, a quintessential compact multiply connected space that incorporates a fifth dimension to encode space-time as a latent manifold.

In some ways, AI is bolder than humans because the huge corpus of knowledge, starting with the prodigious Standard Model (SM) of particle physics, poses almost no burden to its conjecture-framing processes. Why not feed AI with the SM enriched by the troubling cosmological phenomenology on dark matter and dark energy and see where AI takes us vis-à-vis reconciling the conflicting data with the laws of physics? This is precisely the intellectual adventure described in this book and – to the best of our knowledge – in no other book on the shelf. As the reader will discover, many AI conjectures and validations ultimately make a lot of sense, even if their boldness does not feel altogether "human" yet.

This book is written for a broad readership. Prerequisites are minimal, but a background in college math/physics/computer science is desirable. This book does not merely describe what is known about dark matter and dark energy but also provides readers with intellectual tools to engage in a quest for the deepest cosmological mystery.

Chapman & Hall/CRC Artificial Intelligence and Robotics Series

Series Editor: Roman Yampolskiy

Digital Afterlife: Death Matters in a Digital Age
Maggi Savin-Baden and Victoria Mason-Robbie

Multi-UAV Planning and Task Allocation
Yasmina Bestaoui Sebbane

Cunning Machines: Your Pocket Guide to the World of Artificial Intelligence
Jędrzej Osiński

Autonomous Driving and Advanced Driver-Assistance Systems: Applications, Development, Legal Issues, and Testing
Edited by Lentin Joseph and Amit Kumar Mondal

Digital Afterlife and the Spiritual Realm
Maggi Savin-Baden

A First Course in Aerial Robots and Drones
Yasmina Bestaoui Sebbane

AI by Design: A Plan for Living with Artificial Intelligence
Catriona Campbell

The Global Politics of Artificial Intelligence
Edited by Maurizio Tinnirello

Unity in Embedded System Design and Robotics: A Step-by-Step Guide
Ata Jahangir Moshayedi, Amin Kolahdooz, and Liao Liefa

Meaningful Futures with Robots: Designing a New Coexistence
Edited by Judith Dörrenbächer, Marc Hassenzahl, Robin Neuhaus, and Ronda Ringfort-Felner

Topological Dynamics in Metamodel Discovery with Artificial Intelligence: From Biomedical to Cosmological Technologies
Ariel Fernández

A Robotic Framework for the Mobile Manipulator: Theory and Application
Nguyen Van Toan and Phan Bui Khoi

AI in and for Africa: A Humanist Perspective
Susan Brokensha, Eduan Kotzé, and Burgert A. Senekal

Artificial Intelligence on Dark Matter and Dark Energy: Reverse Engineering of the Big Bang
Ariel Fernández

For more information about this series please visit: https://www.routledge.com/Chapman--HallCRC-Artificial-Intelligence-and-Robotics-Series/book-series/ARTILRO

Artificial Intelligence on Dark Matter and Dark Energy

Reverse Engineering of the Big Bang

Ariel Fernández

CRC Press
Taylor & Francis Group
Boca Raton London New York

CRC Press is an imprint of the
Taylor & Francis Group, an **informa** business
A CHAPMAN & HALL BOOK

Cover image: Ariel Fernández

First edition published 2024
by CRC Press
2385 NW Executive Center Drive, Suite 320, Boca Raton FL 33431

and by CRC Press
4 Park Square, Milton Park, Abingdon, Oxon, OX14 4RN

CRC Press is an imprint of Taylor & Francis Group, LLC

ISBN: 978-1-032-46554-8 (hbk)
ISBN: 978-1-032-47404-5 (pbk)
ISBN: 978-1-003-38595-0 (ebk)

DOI: 10.1201/9781003385950

Typeset in Minion
by Deanta Global Publishing Services, Chennai, India

In loving memory of Haydée Stigliano, my mother.

Contents

Preface

THEIR STAGGERING ABUNDANCE AND THE SCALE OF THE ANOMALIES THEY BRING about make it imperative to elucidate the nature of dark matter and dark energy. Unless we get a smarter civilization to whisper the answer to us, it seems that at this point the problem is eminently suitable for artificial intelligence (AI), which does not have to pay respects to physics tradition, is not burdened by the wanton complexities of the Standard Model, and can embark in the boldest assumptions without other restraints than those imposed by logic and consistency.

The AI approach resorts to the underlying science critically revisited in the light of the startling experimental anomalies that arise in deep-space cosmology. As shown in this book, AI is able to illustrate how science is best done: Not by prejudice but through analysis of big data in response to an intriguing hypothesis that humans, too loyal to their scientific cliques, are sometimes unable to formulate or accept.

As we prod the cosmos at very large scales, the pillars of physics seem feebler than ever, just about to crumble under the weight of contradicting evidence. This book helps mitigate the current crisis. It resorts to artificial intelligence for answers and describes the outcome of this quest in terms of an ur-universe, a quintessential compact and multiply connected space that incorporates a fifth dimension to encode Einstein's space-time as a latent manifold.

It turns out that at the largest cosmic scales, a vast amount of matter and movement goes unaccounted, so either there is a colossal surplus of dark matter and dark energy that cannot be detected, or Einstein's theory becomes inadequate to explain deep space. Humans cannot judiciously arbitrate at this juncture, so we must graciously defer to artificial intelligence and let it be the arbiter.

In some ways, AI is bolder than humans because the huge corpus of knowledge, starting with the prodigious Standard Model (SM) of particle physics, poses almost no burden to its conjecture-framing processes. Thus, the plan set forth for this book is to feed AI with the SM enriched with the troubling cosmological phenomenology on dark matter and dark energy and see how AI reconciles the seemingly conflicting data with our currently accepted laws of physics. This is in a nutshell the intellectual adventure that lies ahead for the reader.

Ariel Fernández
North Carolina, USA

Author

Ariel Fernández, PhD, (born Ariel Fernández Stigliano) is an Argentine-American physical chemist and mathematician. He earned a PhD in chemical physics at Yale University in record time and held the Hasselmann Endowed Chair Professorship in Bioengineering at Rice University until his retirement. To date, he has published approximately 500 scientific papers in professional journals, including *PNAS*, *Nature*, *Nature Biotechnology*, *Physical Review Letters*, *Genome Research*, and *Genome Biology*. Dr. Fernández has also authored seven books on biophysics, molecular medicine, AI, and mathematical physics, and holds several patents on technological innovation. Since 2018, Dr. Fernández has headed the Daruma Institute for Applied Intelligence, the research arm of AF Innovation, a consultancy currently based in Argentina and the United States.

CHAPTER 1

Conjuring Up Dark Matter and Dark Energy

Φύσις κρύπτεσθαι φιλεῖ.

– HERACLITUS

("Nature likes to keep its secrets," translation by the author)

SUMMARY

This chapter surveys overwhelming experimental evidence that supports the existence of invisible matter and energy of unknown origin. Dark energy and dark matter contribute significantly to the dynamics of the cosmos at large scales, and their very nature and abundance are likely to trigger a paradigm shift in physics.

1.1 WHY DARK MATTER AND DARK ENERGY?

Physics is an established and respected field of knowledge that endeavors to explain how the universe works at all scales. Its corpus incorporates ideas, models, and data only after careful scrutiny. The bar is high, and scientists who want to leave their mark face a stringent peer review process and only get to impose their views after a hard-won battle. Solid as it seems, there is nothing monolithic, no final word in physics. Truth is never fully conquered but stands as a beacon for the daring. Each time there is a breakthrough, the veil of mystery is lifted a little, but as the horizon expands, new mysteries arise. Every concept, every theory, every measurement is constantly subject to revision as new windows of reality open up to detection and technologies are endlessly perfected. Paradigms, even those that appear rock-hard for a while and endure long-term attrition, often crumble under the weight of new and disconcerting evidence or undergo extensive revision. The history of physics is endlessly made of cycles of destruction and creation, much like the ancient

DOI: 10.1201/9781003385950-1

FIGURE 1.1 Shiva as "Nataraja," the lord of dance, the destructive deity that also dictates the rhythm of the universe. Credit: Los Angeles County Museum of Art. Image from the public domain (https://en.wikipedia.org/wiki/File:Shiva_as_the_Lord_of_Dance_LACMA_edit.jpg).

cosmogonies of the valley of the Indus, dictated by the infinite toils of Shiva, the destructive element that dances out the pulse of the universe (Figure 1.1).

In or around 1900, the Irish physicist William Thomson (1824–1907), usually referred to as Lord Kelvin, famously declared: "There is nothing new to be discovered in physics now. All that remains is more and more precise measurement." In the two or three decades that followed this pronouncement, two earth-shattering revolutions in physics took place: Einstein's theory of relativity and quantum mechanics [1]. So much for Lord Kelvin's solemn pronouncement… History proved once again to be the master of irony.

Relativity and quantum physics thrived and prospered because they effectively and successfully addressed shortcomings in the prevailing paradigm at the turn of the 20th century. This paradigm is essentially enshrined in two basic pillars of knowledge: (a) Newton's law of gravitation, which governs the dynamics of falling and orbiting bodies at terrestrial (the apocryphal falling apple) and cosmic scales, and (b) Maxwell's laws governing electromagnetic phenomena, that is, the events that reveal the entanglement between electricity and magnetism [1]. It thus seemed that all the effects involving the known forces in the universe at the time were satisfactorily understood, even if the nature of such forces remained unyielding to theoretical efforts. For example, when asked about the nature of gravity, Newton snapped in Latin: "*Hypotheses non fingo*" (I contrive no hypothesis). In an ironic turn, right after Lord Kelvin's pronouncement, Einstein came up with relativity, the first theory that truly explained gravity, while an avalanche of new data revealed that matter at atomic and subatomic scales required a drastic revision of the extant conceptual framework, heralding the birth of quantum mechanics.

Einstein showed that time and space are inevitably entangled and should not be treated separately, with time acting as the fourth dimension. His view of gravity in "spacetime" is admirably synthesized in the quote of the American physicist John Archibald Wheeler (1911–2008): "Spacetime tells matter how to move, matter tells spacetime how to curve" [1]. On the other hand, quantum mechanics successfully accounted for the discontinuous nature of energy and momentum experimentally shown to hold when the behavior of matter is observed at atomic and subatomic scales. Thus changes in those physical magnitudes are often accompanied by emission or absorption of radiation, which can only vary in a discrete fashion, as multiples of a constant. These packages of discontinuous energy are known as quanta, a term coined reluctantly, as it turns out, by the German physicist Max Planck (1858–1947).

Lofty and sturdy as it may seem, the edifice of contemporary physics is beginning to show major cracks and may not withstand the attrition to which it is exposed. The cracks – we now know – are not superficial but extend all the way to the very foundations. As the dynamic structure of the universe is examined at extremely large scales, commensurate with the dimensions of galaxies, and, further, of clusters and other assemblages of galaxies, major anomalies are surfacing. Make no mistake: Legions of physicists are losing their sleep over the problem. In fact, the anomalies have been surfacing since the 1930s, although the conceptual cracks were initially patched up with soft filling. Now we know the cracks are indeed structural and require immediate attention, or the whole edifice of physics, the crowning achievement of human civilization, may be doomed and perhaps – let us add some melodrama at this point – even tagged for a paradigm demolition.

Numbers simply don't add up in the cosmos. In the outer shells, stars in spiral galaxies have been behaving in ways that can be considered anomalous – to use physicists' typical euphemism – spinning at speeds far larger than those that would enable gravity from the visible universe to hold them in stable orbits. From this perspective, for the universe to make sense, the gravitational pull must be far larger than what is expected from the amount of matter detected [1, 2]. This begs the question: Where is the missing matter? Or are we supposed to introduce a fudge term in the equations? This disquieting picture of galaxy rotation became apparent through the pioneering work of the American astronomer Vera Rubin (1928–2016). Vera examined six spiral galaxies, like our Milky Way, or nearby Andromeda, and consistently found outer stars behaving anomalously: There is simply not enough detectable mass to keep them in their orbits. Newton's law of universal gravitation worked astonishingly well when it comes to describing the dynamics of our solar system – which earned Newton a superlative reputation – but when it comes to stars revolving around the center of galaxies, we are dismayed by the outcomes of the theory. We are left with two painful alternatives: Either the law of universal gravitation fails miserably or there is a huge amount of matter out there that we cannot detect and introduces a considerable gravitational pull, five times larger than ordinary matter [1, 2].

This is not the only problem with the cosmos – not even the worst one – that physicists are currently losing their sleep over: The universe is expanding at an ever-increasing rate that far exceeds what the gravitational pull would enable and far exceeds the amount of kinetic energy available from the motion of visible matter. It seems that plenty of the matter required to sustain the structure and dynamics of the universe in accord with Newtonian or even relativistic laws simply cannot be accounted for. It is simply invisible. The missing matter was named "dunkle Materie" (dark matter) by the Swiss astronomer Fritz Zwicky (1898–1974), who first postulated its existence, although perhaps "invisible matter" would have been a better name. Aside from his brilliance as an astrophysicist, Zwicky seems to have been quite a colorful character, a curmudgeon, and a "lone wolf" by his own admittance. His intrepid forays into the heavens gave new impetus to astrophysical research [3].

Equally or even more troublesome, the rate of expansion of the universe is not even constant: It is going at an exponential rate (Figure 1.2). As their speed exceeds the speed of light, zillions of stars will simply spin out of our horizon and become forever undetectable: Their light will never reach us. For this staggering discovery, astrophysicists Saul Perlmutter, Brian Schmidt, and Adam Riess were awarded the 2011 Nobel Prize in physics [1, 3]. What is the source of this enormous surplus in kinetic energy seemingly sprouting from the vacuum of deep space? Nobody is certain, although a quantum mechanical origin is often invoked, hence the term "quantum vacuum fluctuations." A description of this mysterious energy gets too technical and is deferred to Chapters 3 and 6. This view postulating quantum vacuum energy is problematic, to say the least, since the naïve energy estimations are off by a factor of 10^{120} (1 followed by 120 zeros) when contrasted against the experimental

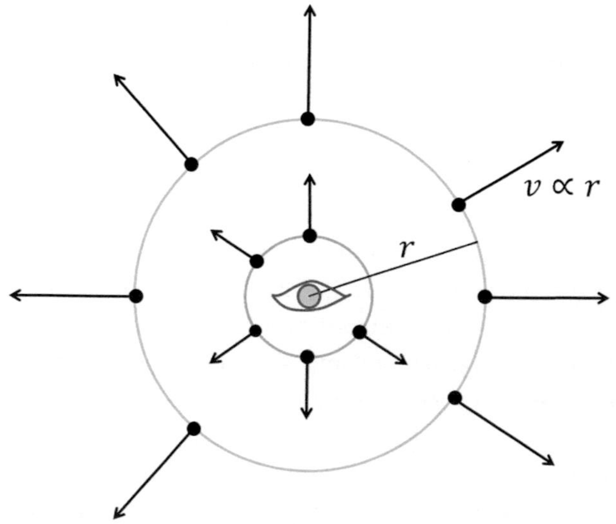

FIGURE 1.2 Schematic representation of a universe undergoing accelerated expansion. The component of the velocity of a celestial object contributing to its getting away from the observer increases as the object is farther away from the observer. Distant objects (larger radius *r*) have been traveling for longer times, and hence their speed has had a chance to increase more relative to those that have traveled less.

results obtained by Perlmutter, Schmidt, and Riess! The surplus kinetic energy associated with the berserk expansion of the universe has been named "dark energy."

Dark matter and dark energy cannot be considered corrections to the laws that presumably govern the cosmos [2, 3]. For starters, there is roughly five times more matter in the universe than visible matter: The matter we are able to detect. We know this from the gravitational influence of dark matter on visible matter. Furthermore, we know that energy and matter are interchangeable, with one becoming a proxy for the other and vice-versa, as implied by the arch-famous Einstein equation $E=mc^2$ (E=energy, m=mass, c=speed of light). With this formula in mind, the breakdown of the actual mass budget of the universe looks even weirder. Give and take a few tens of a percent, current state-of-the-art calculations yield the following composition of the universe:

Ordinary matter: 5.0%

Dark matter: 26.7%

Dark energy: 68.3%

These proportions are nothing short of scandalous. We need to come to grips with the fact that the label dark matter simply captures our ignorance regarding the nature of most of the matter in the universe. Shockingly, 95% of the universe may be accounted for but remains utterly undetectable. We infer its existence indirectly, through its gravitational influence on ordinary matter.

Invisible matter now present at temperatures almost negligible (3K, or −454°F, or −270°C) originated at much higher temperatures prior to the formation of galaxies, yielding the name "cold dark matter" (CDM) from the fact that it moves at non-relativistic speeds ($v<<c$) [2]. This material of unknown nature first accrued into small galaxies and subsequently served as building-block and seeding or nucleating material for larger scale structures up to the present-day gravitationally bound clusters of galaxies. In the widely accepted standard cosmology model, the gravitational growth of present-day galaxies and their clustering is steered by primeval fluctuations that animated a sea of cold dark matter. Although the nature of dark matter is anyone's guess at this point, astrophysicists have measured the imprint of their fluctuations in their primeval spatial distribution.

This may begin to sound like science fiction, but such ancient imprint is embossed as slight variations across the universe in the brightness of the so-called cosmic microwave background (CMB), the relic ultra-weak radiation field (Figure 1.3) left over from the Big Bang [1]. The CMB is the landmark evidence of the Big Bang origin of the universe and constitutes a faint radiation that fills up all space, dating back to the time when atoms were first formed. With optical telescopes, the space in the background of light-emitting objects is completely dark. Only a very sensitive radio telescope shows a faint background noise, a glow not associated with any object in the sky, a signal that is strongest in the microwave region of the spectrum. Its accidental discovery in 1965 is credited to American astronomers Arno Penzias (1933–) and Robert Wilson (1936–), earning them the 1978 Nobel Prize in physics [1]. In colloquial terms, we may say, we are still "hearing" that massive explosion

FIGURE 1.3 Cosmic microwave background (CMB) displaying variations in the intensity of the radiation. The CMB is a snapshot of the oldest light in our universe, imprinted on the sky when the universe was a mere 380,000 years old. The signal is faint and lies in the radio wave spectrum. It shows tiny temperature fluctuations that correspond to regions of slightly different densities, representing the seeds of all future structures in the universe: The stars and galaxies of today. Credit: US National Aerospace Agency (NASA). Image from the public domain (https://commons.wikimedia.org/wiki/File:Ilc_9yr_moll4096.png).

as the CMB (the radiation is in the radio wave frequency range). Spatial changes in the intensity of what we are hearing may give us clues on how dark matter got organized and distributed after the big cool down that followed the Big Bang. The acoustic oscillations detected experimentally to exquisite precision in the brightness fluctuation smudges of the CMB indicate the presence of a dominant invisible form of matter that flows freely, noble and inert, alongside ordinary matter and radiation that are tightly coupled through electromagnetic interactions.

Today, many experiments are frantically searching for signatures of dark matter, both in the sky and in the laboratory, including the widely known Large Hadron Collider (LHC), a massive international consortium built near Geneva, Switzerland, to discover and detect subatomic particles [2]. This search has so far been unsuccessful, and this book does not conceal some skepticism regarding the outcome of LHC experiments in regard to dark matter, as discussed in subsequent chapters.

As suggested by the path-breaking work of Vera Rubin, the revolution dynamics of stars and dust in galaxies imply the existence of invisible mass in a halo that extends well outside the inner region where ordinary matter concentrates [1, 3]. Surprisingly, the need for dark matter in galaxies appears only in the outer region where the gravitational acceleration drops below a universal value, which equals roughly the speed of light (299,792.458 kilometers or roughly 186,000 miles per second) divided by the age of the universe (13.82 billion years = 436,117,077,000,000,000 seconds). This is a highly disconcerting fact within the favored interpretations of dark matter. The sheer existence of a universal threshold (0.0000000007 meters per second squared) in the acceleration due to gravitation raises the daunting possibility that we are not actually missing matter but rather witnessing a change in the effect of gravity on the dynamics of visible matter at extremely low values of the gravitational force. This would require a very significant revision of the cosmology models enshrined in the theories of Newton and Einstein, a revision that may be tantamount to a dramatic paradigm change, bringing shockwaves to the scientific community. We had better stick to dark matter, for which there is evidence now coming from a variety of different sources…

1.2 QUERYING AI ON DARK MATTER AND DARK ENERGY

More definitive clues are needed to figure out the nature of dark matter and dark energy. This book squarely addresses this imperative. In their quest, humans have at their disposal and also carry the burden of an enormous corpus of knowledge across a variety of fields, from particle physics to cosmology. Thus, a solid command of the so-called Standard Model in particle physics may surely become a blessing, as it provides a foundational substrate to build upon, but in some sense, it may also become a handicap since the sheer volume of knowledge may hamper the boldness of approach that the dark matter/dark energy conundrum demands. Unless we get a smarter civilization to whisper the answer to us, it seems that at this point the problem is eminently suitable for artificial intelligence, which does not have to pay respects to tradition and can embark on the boldest assumptions without other restraints than those imposed by logic and consistency. This is precisely the approach adopted and described in this book in the most elementary way possible for the benefit of a broad audience.

The approach resorts to the underlying science critically revisited by artificial intelligence in light of the startling anomalies observed experimentally. When properly steered, artificial intelligence illustrates how science is best done: Not by prejudice but through analysis of new big data in response to an intriguing hypothesis that humans, too loyal to their scientific cliques, are sometimes unable to formulate or accept.

As argued previously, as we prod the cosmos at very large scales, basic tenets of physics seem to crumble under the weight of contradicting evidence. This book helps mitigate the current crisis. It resorts to artificial intelligence for answers and describes the outcome of this quest in terms of an ur-universe, a quintessentially compact and multiply connected space that incorporates a fifth dimension to encode space-time as a latent manifold.

The American physicist John A. Wheeler aptly characterized Einstein's universe with the phrase: "Matter tells space how to bend while space tells matter how to move." But it turns out that at the largest cosmic scales, plenty – the vast majority – of matter and movement goes unaccounted, so either there is a colossal surplus of dark matter and dark energy that cannot be detected, or Einstein's theory becomes inadequate to explain deep space. Humans cannot judiciously decide at this juncture, so we must let artificial intelligence be the arbiter.

In some ways, AI is bolder than humans because the huge corpus of knowledge, starting with the prodigious Standard Model (SM) of particle physics, poses almost no burden to its conjecture-framing processes. So, the plan set forth for the rest of the book is to feed AI with the SM, enriched with the troubling cosmological phenomenology of dark matter and dark energy, and see how AI reconciles the seemingly conflicting data with the currently accepted laws of physics. This is in a nutshell the intellectual adventure that lies ahead for the reader.

REFERENCES

1. Weinberg S (2008) *Cosmology*. Oxford University Press, New York
2. Fisher P (2022) *What Is Dark Matter?* Princeton University Press, Princeton, NJ
3. Clegg B (2019) *Dark Matter and Dark Energy: The Hidden 95% of the Universe*. Icon Books, London, UK

CHAPTER 2

Dark Matter in Galaxies

Coming to Grips with an Inevitable Truth

To a man with a hammer
everything looks like a nail.

– MARK TWAIN

SUMMARY

Leveraging fairly elementary physics arguments, we examine experimental data on star rotation in galaxies, where anomalies in the rotation speed lead us to conjecture the existence of dark matter. This is a form of invisible matter of unknown origin that significantly enhances the gravitational field of galaxies, thereby generating the required centripetal force to keep stars in their orbits.

2.1 DARK MATTER IN GALAXIES AND ANOMALIES IN STAR ROTATION

As keen observers open to the mysteries of the cosmos, we are now taking a closer look at dark matter. Of course, we cannot expect to see or detect dark matter directly, but we must be prepared to infer its existence from its influence on ordinary (visible) matter. This influence is known to be solely gravitational, as dark matter does not appear to be implicated with other forces of nature. Essentially, we shall strive to give physical reasons why we think that there is such a thing in the first place. Let us consider the best understood and simplest possible celestial motion, i.e., that of a planet orbiting around a star. For simplicity, the motion may be assumed to be circular, with the star at the center. Thus, the planet moves in its orbit with velocity v and is located at all times at distance r from the star (Figure 2.1). For the sake of illustration, we assume the star is our sun and we are describing planetary orbits in our solar system. This picture is only an approximation, as planets typically describe elliptical orbits with the sun in one of the foci of the ellipse (Figure 2.1). A key physical quantity in this picture is the *centripetal force* – let's name

DOI: 10.1201/9781003385950-2

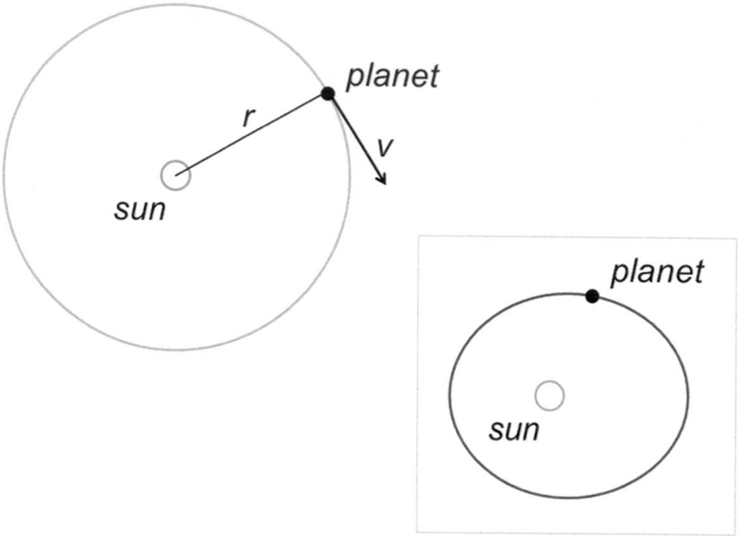

FIGURE 2.1 Scheme of a planet orbiting around the sun. The circular orbit is an approximation. In reality, most planets describe elliptic orbits (inset).

it F – that is, the force toward the center of the circular motion provided by the gravitational pull of the sun and required to keep the planet in its stable circular orbit. This force was first calculated by Newton from the simple and rigorously obtained relation $F=mv^2/r$, where m is the mass of the planet, v is its velocity along its circular orbit, and r is the – constant – distance between the orbiting planet and the sun or, equivalently, the radius of the circular orbit, whichever the reader prefers. As said, this force is provided by the gravitational pull of the sun, hence it is equal to the gravitational force, which according to Newton's law of universal gravitation is given by $F=GMm/r^2$, where M is the mass of the sun and G is Newton's gravitational constant [1]. By combining both expressions of the centripetal force we get

$$F = \frac{mv^2}{r} = \frac{GMm}{r^2} \tag{2.1}$$

Equation (2.1) yields $v^2=GM/r$, or $v = \sqrt{GM/r}$. This result is very important to describe the solar system since it implies that the further away a planet is from the sun, the slower the speed at which it revolves around the sun [1]. We can qualitatively plot this planetary behavior for the solar system as shown in Figure 2.2.

Now let us consider a much larger scale in the cosmos. Instead of planetary systems, let us consider galaxy rotation, or more precisely, the dynamics of stars that revolve around the center of spiral galaxies. We would imagine that this motion would be similar to that of planets revolving around stars, but that is not the case: Fundamental anomalies arise, and – as it turns out – these anomalies are key to inferring the existence of dark matter.

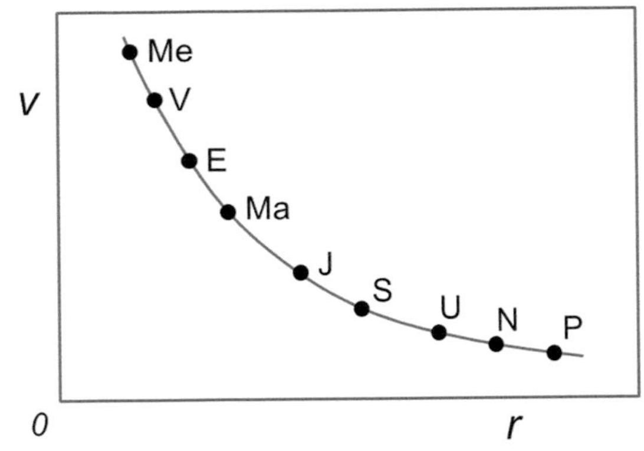

FIGURE 2.2 Qualitative scheme of the correlation between velocity and distance to the sun for the nine planets of the solar system (Me=Mercury, V=Venus, E=Earth, Ma=Mars, J=Jupiter, S=Saturn, U=Uranus, N=Neptune, and P=Pluto).

Since the visible mass M of the galaxy exerting a gravitational pull on the star is mostly concentrated at and "near" the center of the galaxy, we are likely to expect a v–r correlation similar to that presented in Figure 2.2, with v being now the speed of the star, and r being the distance to the center of the galaxy. In this case, our prediction would extend to "short" galactic distances, where the mass contained inside the radius of the star orbit, the mass that exerts a gravitational pull on the star, becomes significantly smaller as we approach the center of the galaxy, as more and more mass is left outside the star orbit. Based on experimental observation of visible matter in the heavens, we are assuming that the mass is concentrated at and near the center of the galaxy and that, as we navigate toward the boundaries of the galaxy, mass concentration is negligibly small relative to the concentration near the center.

Like in planetary motion, the behavior of the revolving velocity of the star is predicted again to be $v^2=GM/r$, except that now M decreases substantially for small r, as less visible matter lies inside the star orbit with decreasing r. Let us get specific at this point: The detailed prediction for spiral galaxy Messier 33 is presented in Figure 2.3. On the other hand, experimental work paints a very different picture: Observation of the galaxy following the pioneering work of Vera Rubin reveals a very different behavior, particularly for distant or outer stars in the galaxy, as shown in Figure 2.4. The measured velocities of the outer stars are seemingly less dependent on the distance to the center of the galaxy [1]. For outer stars, as the distance r to the center of the galaxy increases, M, the mass contained within the orbit of the star, hardly changes at all since, as we have said, most of the mass is concentrated at or near the center of the galaxy. Yet the velocity of the star is no longer decreasing according to the Newtonian formula $v^2=GM/r$ (Figure 2.4). The experimental measurements shown in Figure 2.4 have been corroborated time and again by independent observers around the world and are strikingly similar for all examined spiral galaxies [1]. The results invariably show that outer stars in galaxies are behaving anomalously: They are spinning so fast that the centripetal force or gravitational pull exerted by the galaxy would not be enough to keep them in orbit, yet the stars are clearly revolving around the center

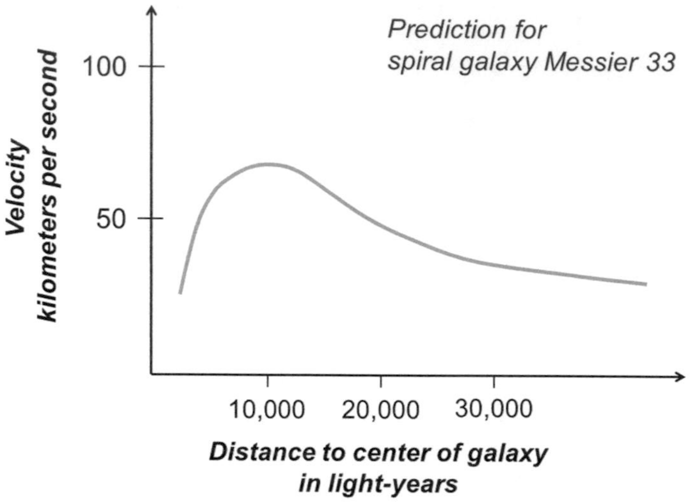

FIGURE 2.3 Predicted correlation between the velocity of a star revolving around the center of the galaxy and its distance to the center of the galaxy. The curve was obtained for the spiral galaxy Messier 33.

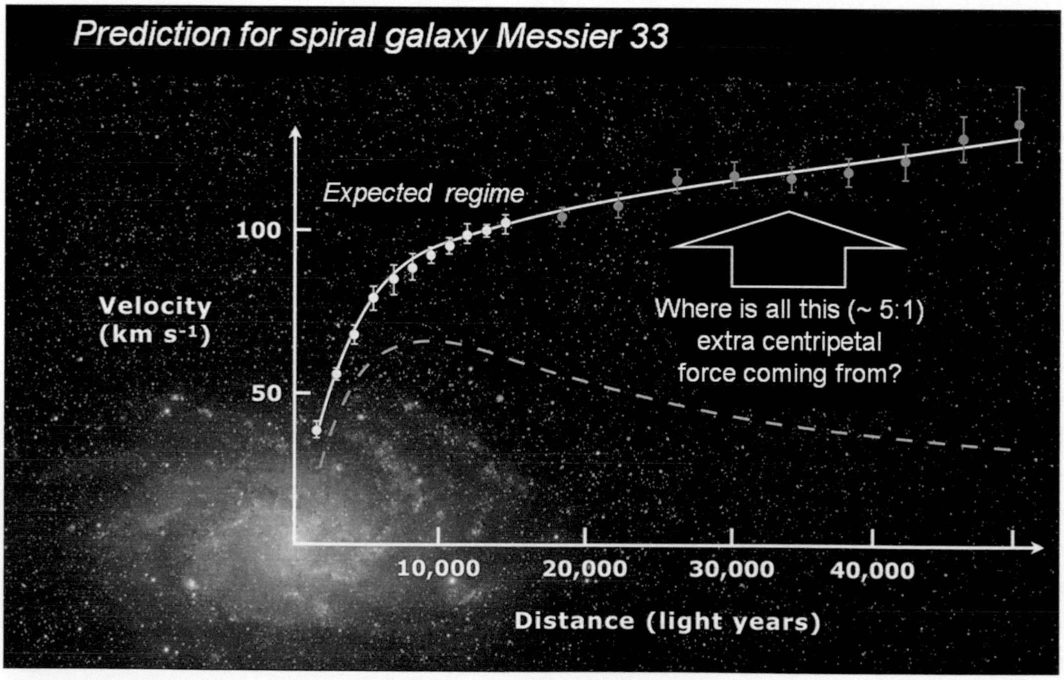

FIGURE 2.4 Rotation curve for spiral galaxy Messier 33 (points with error bars) and predicted curve from distribution of the visible matter (dashed gray line). The prediction and experimental measurement give the velocity of star rotation around the center plotted against the distance of the revolving star to the center of the galaxy. The huge discrepancy between theoretical and experimental values is attributed to dark matter distributed in a halo surrounding the galaxy. Adapted from a figure in the public domain. Credit: Mario De Leo – own work, CC BY-SA 4.0 (https://commons.wikimedia.org/w/index.php?curid=74398525).

in stable orbits. According to Newtonian physics, at the measured speeds the outer stars in the galaxies should fly away into outer space. This is clearly *not* what is happening.

Taken together, these observations clearly and unambiguously suggest that Newtonian physics breaks down; it is no longer upheld. The only alternative that would bring some intellectual relief to the physics community is that the mass M of the galaxy, responsible for the gravitational pull on the star, has been grossly underestimated. But how could that be? All detectable matter has been detected and the mass computed with satisfactorily convergent results from constantly perfected technologies. These anomalies prompted physicists to postulate the existence of dark matter, which is undetectable matter of an unknown nature distributed as a halo around the galaxy. This type of matter would contribute significantly to the gravitational pull necessary to ensure the stability of the orbits of outer stars.

Thus, the only inevitable conclusions from results such as those shown in Figure 2.4 are that either a) the physics that withstood centuries of attrition needs a major correction, or b) the mass M in the gravitational equation that governs the motion of the stars in the galaxy represents something else, something quite different from the mass of the ordinary (detected) matter. In either case, both assumptions would represent fundamental departures from what we have been taking for granted for the last 300 years.

We could in principle modify the sacred Newton's gravitational law $F=GMm/r^2$ for an extremely large distance r, when the gravitational pull is very tenuous. After all, Newton never dealt with colossal distances of the order of tens of thousands of light years (one light year is approximately 5.88 trillion miles), like those represented in Figure 2.4. The extant technology in the 17th century did not enable such observations. In fact, data regression analysis of experimental results of the type presented in Figure 2.4 has led to a modification of Newton's law adapted for huge intra-galactic distances. Thus, the Israeli physicist Mordehai Milgrom dared to modify Newton's equation in an effort to adapt it to the outpour of astrophysical data. While commendable, his effort is mostly regarded as a formula for data regression, essentially a data-fitting device and not a fundamental advance in the physics underlying galaxy rotation [1].

The physics community overwhelmingly favors the dark matter hypothesis and has chosen to uphold the sanctity of Newton's law [2]. As we shall now see, there are other experiments that support the dark matter hypothesis and favor it over Milgrom's defiance of Newtonian physics [1, 2].

2.2 GRAVITATIONAL LENSES AS DARK MATTER DETECTORS

There is a well-known way of determining the mass of a galaxy, which is actually a feature of Einstein's general relativity called "gravitational lensing." As mentioned in the previous chapter, a massive object such as a galaxy will bend or warp the fabric of space-time, causing a change in curvature, in the form of a dimple (Figure 2.5). Since light travels in space-time, it will be affected by the gravitational pull exerted by the massive object. This interaction between gravity and light elicited huge skepticism when first proposed by Einstein but ultimately proved to be a decisive validation of general relativity as gravitational lenses were experimentally confirmed. Before dwelling on the mechanism of this phenomenon and leveraging this relativistic feature as a means to detect dark matter, we

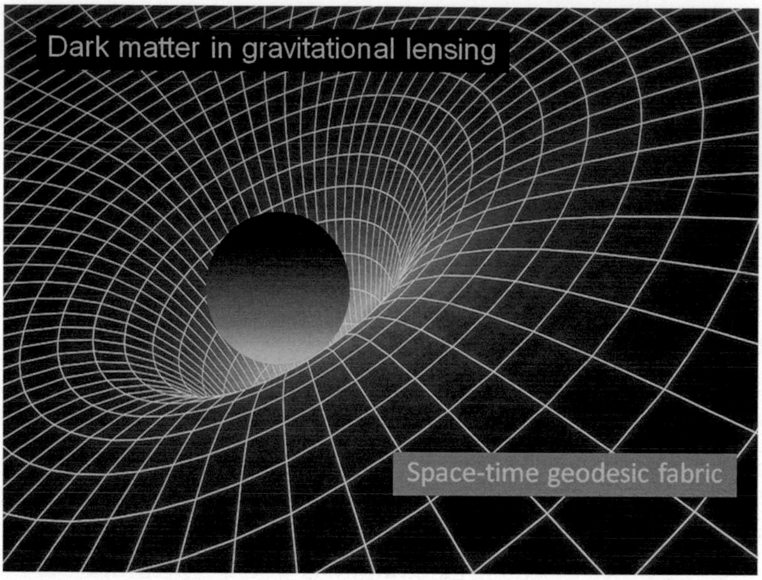

FIGURE 2.5 Relativistic picture of space-time (grid plane) curved or warped by the presence of a massive object.

need to prove rigorously that light may be indeed influenced by gravity. This amazing observation is not in the least obvious since the photons, the "particles" of light, are known to be massless.

2.2.1 Why Did Einstein Claim That Gravity Influences Light?

This question is at the core of relativity. To show that light is indeed affected by a gravitational field, we shall resort to Einstein's favorite theoretical tool, the thought experiment (*Gedankenexperiment* in German). No fancy equipment is required, no expensive gadgets, only the imagination, unbound, tempered only by the principles of physics and their logical consistency. In fact, let us play Einstein for a while. We can imagine young Albert at his modest frugal desk in the patent office in Bern, eyes closed, complete silence, prodding his imagination as he posits his favorite question "What if…?"

Let us imagine a box floating in deep space (Figure 2.6); no forces are acting on it. A photon of light leaves one of the walls of the box and travels to the opposite wall on the other side of the box. Let us place the center of coordinates at the point on the wall where the photon started its trajectory. As the photon is absorbed, it transfers some modicum of energy that we may denote E (leave your car in the sun and you will have trouble getting back in it). This implies that the photon must carry momentum, that is, what we would identify with "impetus" in common parlance. But the momentum, p, of a particle is usually assessed as $p=mv$, where m denotes the mass of the particle and v is its velocity. Since we are told that the photon is massless, how can it possibly have momentum? This question troubled young Einstein for a while. The photon travels at the speed of light, $v=c$, so to compute the momentum of the photon, Einstein used the equation $p=E/c$. This equation

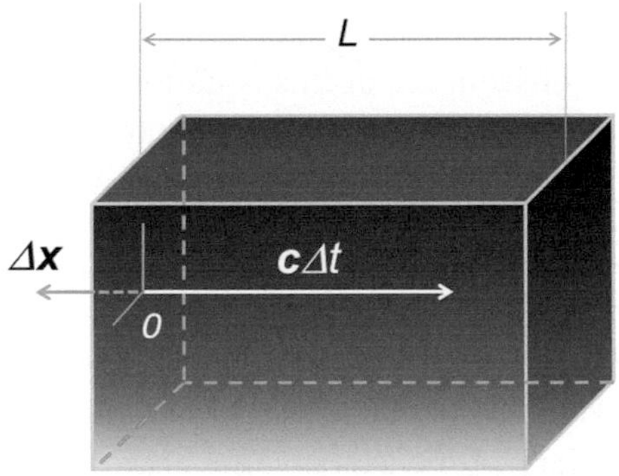

FIGURE 2.6 Representation of the thought experiment that led Albert Einstein to formulate his famous equation $E=mc^2$.

does not involve the mass, the troublemaker. As it turned out, that was a clever move to circumvent the problematic mass of the photon, since Einstein knew that the kinetic energy of a moving object traveling at speed v may be computed as $E=mv^2=pv$, so Einstein simply wrote $p=E/v$, which becomes $p=E/c$ for the photon that travels at the speed of light. This turned out to be a great trick!

Like most of us, Einstein was also familiar with the principle of action–reaction. Everyone who has fired a gun knows this principle: When the gun fires, the bullet goes one way, the gun the other – it's commonly known as recoil or kick. This principle has a fancy name in physics, it is known as the principle of momentum conservation. When we apply it in the context of our thought experiment, we need to equate the reaction of the box to the movement of the photon inside. Both momenta must be equal in magnitude to compensate for each other. Thus, at time Δt from the moment the photon left the wall, the following relation must hold:

$$\frac{M\Delta x}{\Delta t} = \frac{E}{c}, \tag{2.2}$$

where $\Delta x/\Delta t$ is the velocity of the box "recoiling" from the photon departure, M is the mass of the box, and Δx is the distance traveled by the box at time Δt, while in that time the photon traveled a distance $c\Delta t$ in the opposite direction (Figure 2.6).

Since the photon travels at the speed of light, the time it takes for it to reach the opposite wall of the box at distance L is $\Delta t=L/c$. Now suppose we replace the photon with a particle with mass. At the time $\Delta t=L/c$, the conservation of momentum would read: $M\Delta x/\Delta t = mL/\Delta t$, or $M\Delta x = mL$. Using Equation (2.2), we can determine that at time $\Delta t=L/c$, the box has been displaced by the amount $\Delta x = E\Delta t/Mc = EL/Mc^2$. Since we already showed that $M\Delta x = mL$, we get $MEL/Mc^2 = mL$, which is rewritten as $E=mc^2$, Einstein's arch-famous equation!

This means that mass and energy are interchangeable, a result we shall exploit to investigate the nature of dark matter and dark energy. As is well known, since c^2 is a colossally large quantity, it transpires that a small amount of mass is capable of transmutation into a huge amount of energy. This observation heralded the power of nuclear energy in a weaponizing context. Another consequence of this equation of direct relevance to our previous discussion of gravitational lensing is that since a particle with mass $m = E/c^2$ may be regarded as a proxy for a photon carrying energy E, light is indeed affected by gravitation, as Einstein correctly predicted.

2.2.2 The Physics of Gravitational Lensing as a Dark Matter Detector

To understand the principle, let us consider a photon of light (the minimal package of light) that barely skims the boundary of the galaxy. This beam will be deflected at an angle α (Figure 2.7). A general relativity calculation of the deflection of light caused by the gravitational influence of the galaxy gives the rigorous result:

$$\alpha = 4MG / Rc^2 \tag{2.3}$$

In the equation, M is the total mass of the galaxy, and R is the radius of the galaxy.

We can also grasp this phenomenon and derive the formula from Equation (2.3) from basic physics. Newton assumed that particles were corpuscles, perhaps not endowed with mass but certainly with momentum. We now know that the photon has momentum (E/c) but no mass, yet from the relation $E=mc^2$, we may assume that it is influenced by gravity

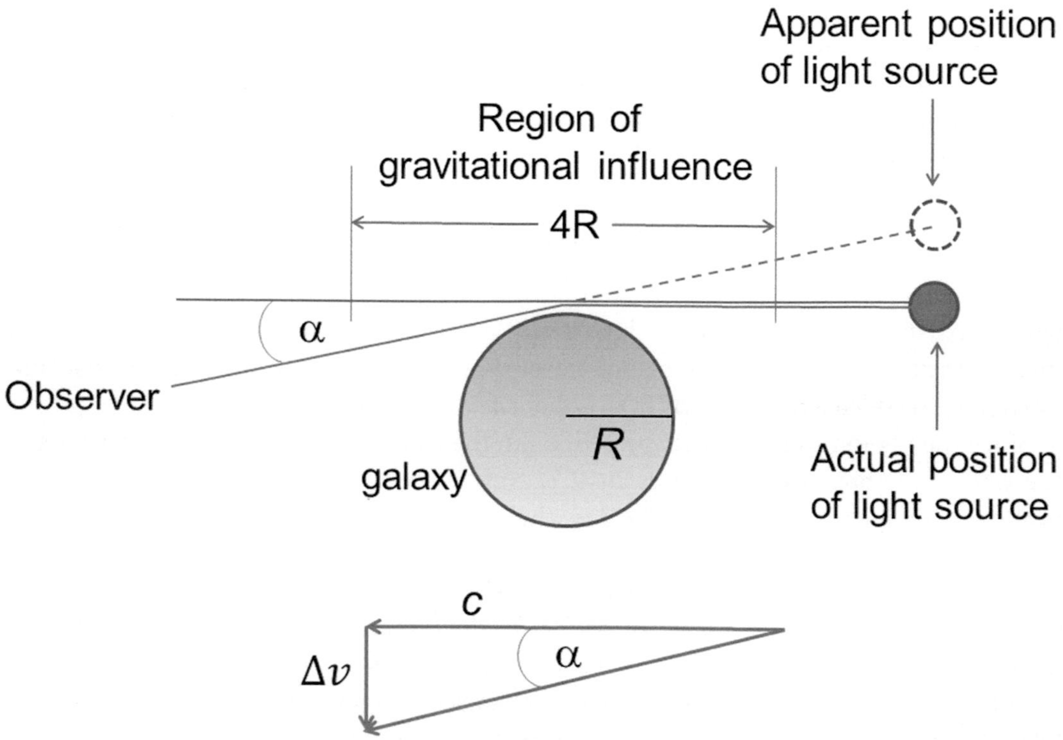

FIGURE 2.7 Elements of the physics of gravitational lensing.

since mass becomes a proxy for energy, just like Einstein predicted. Adopting this picture, the photon with velocity c (the speed of light) would undergo a deflection of its straight-line trajectory and hence its velocity vector (arrow of magnitude *c*) will change direction under the gravitational influence of a massive object. This change in direction requires a force; that is, the photon is subject to acceleration, *a*, perpendicular to its original direction. The force that deflects the light is simply the familiar $F = ma = GMm/R^2$, where *M* is the mass and *R* is the radius of the galaxy. This means that the acceleration undergone by the photon of light as it passes by a galaxy is $a = GM/R^2$. Since acceleration is the change in velocity over time, that means that its change in velocity Δv would be given by $\Delta v = a\Delta t$, where Δt is the time interval during which the galaxy exerts a significant gravitational pull. We may assume this force is maximized as light is closest to the galaxy, and since the galaxy radius is *R*, the force is fully exerted during the traveled distance 4*R*, as indicated in Figure 2.7. Note that 4*R* is an enormous distance, of the order of tens of thousands of light years, as illustrated in Figure 2.4. But the photon travels at the speed of light so that the time interval becomes $\Delta t = 4R/c$. This means that the perpendicular change in velocity caused by the gravitational pull of the galaxy is $\Delta v = a\Delta t = \frac{(GM/R^2)4R}{c} = 4MG/Rc$. So, assuming the deflection angle α is small, we may use the approximation

$$\alpha \approx sin\alpha = \frac{\Delta v}{c} = 4MG/Rc^2. \tag{2.4}$$

The formula in Equation (2.4), based on crude assumptions and obtained from basic physics, is *identical* to the one obtained by general relativity (Equation (2.3)).

This result demonstrates the astonishing consistency between general relativity and Newtonian physics.

The deflection angle in gravitational lensing of massive galaxies has been measured experimentally, and it is approximately six times larger than the value obtained from general relativity or classical physics. *Because the deflection angle is proportional to the total mass* M *of the galaxy, this result implies that the total mass* M *of the galaxy is about six times larger than the value calculated from the visible and detected matter in the region contained within the sphere of radius* R, *the radius of the galaxy*. This is a shocking result and implies that the matter that cannot be detected; dark matter contributes five times more than visible matter to the gravitational pull of the galaxy [2]. This result is in striking agreement with the proportion of dark matter to visible matter in the universe given in Chapter 1: (26.7):5 ≈ 5.3.

2.3 THE DARK MATTER HYPOTHESIS IS UPHELD

This chapter illustrated the way physics works best. Experimental observations that stand in defiance of a prevailing paradigm are initially invariably regarded as "anomalies" or "systematic errors," and only if they pass the initial peerage scrutiny, they may be taken seriously as worthy contenders of an existing paradigm. This implies that the standard model or laws that are expected to underpin the newly observed phenomena may undergo revision or be completely reformulated in a new guise that often represents a synthesis of two clashing proposals of reality. This narrative describes the saga of dark matter as it unfolded in this chapter. The extant physical models and laws cannot be adapted to encompass the experimental observations in the contexts of galaxy rotation and gravitational lenses unless the mass responsible for the gravitational pull gets a significant contribution from matter of a hitherto unknown nature. This is precisely dark matter, and only an approach bold enough to be unencumbered from the weight of standard models can delineate its nature. As we shall show later in this book, the time seems to be ripe for AI to prod over the mass astrophysical data and elucidate the nature of dark matter.

REFERENCES

1. Weinberg S (2008) *Cosmology.* Oxford University Press, New York
2. Fisher P (2022) *What Is Dark Matter?* Princeton University Press, Princeton, NJ

CHAPTER 3

Dark Energy Is Fueling a Runaway Universe

Suppose someone were to say "Imagine this butterfly exactly as it is, but ugly instead of beautiful."

– LUDWIG WITTGENSTEIN

SUMMARY

We describe experimental cosmological evidence on the accelerated expansion of the universe that supports the existence of dark energy. This is a form of energy of unknown origin that gets created through an autocatalytic mechanism and thus fuels the universe's runaway process.

3.1 AN EXPANDING UNIVERSE IN THE CLASSICAL NEWTONIAN WORLD: THE TROUBLED ORIGIN OF A RELATIVISTIC EQUATION

Evidence that massive amounts of energy of unknown origin keep pouring into the universe without undergoing dilution has been piling up since 1998. That was the year astronomers verified the accelerated expansion of the universe. As space expands, more vacuum forms, and with it more geometrically undiluted energy is infused into space, causing space to expand even further in a sort of autocatalytic reaction, where dark matter creation appears to be self-stimulated.

The argument for accelerated expansion runs as follows: Expansion of space stretches light, shifting it to longer wavelengths, hence shifting light toward the red section of the visible spectrum. Light from supernovae appears more "redshifted" the farther away they are from us because their light has to travel farther through an expanding space. If space expanded at a constant rate, a supernova's redshift would be proportional to its distance to the observer and thus to its brightness. This is not what astronomers have been observing.

DOI: 10.1201/9781003385950-3

In an accelerating universe filled with undiluted dark energy that scales with vacuum dimensions, space expanded more slowly in the past than it does now. This means a supernova's light will have stretched less during its journey to Earth, given how slowly space expanded during much of the time compared with the speed at which it expands now. The light from a supernova located at a given distance away (as determined by its brightness) will appear significantly more redshifted than it would in a universe that lacks dark energy. Indeed, researchers find that the redshift and brightness of supernovae scale in precisely that way. The actual scaling enabled them to compute the amount of dark energy in the universe with significant precision.

At the turn of the 20th century, the cosmological debate had a very different flair. In order to calibrate his ideas as he started developing his general theory of relativity, young Einstein had many conversations with astronomers. They assured him that the universe was essentially static, with no beginning and no end, with galaxies at fixed positions unchanged on a cosmic scale. This picture posed a great problem to Einstein as he sought to reconcile it with the idea that galaxies exerted a gravitational pull on one another that would ultimately lead to a collapse of space-time into a big crunch. Einstein circumvented this problem by introducing a "cosmological constant," a fudge term representing a force that would counter the gravitational pull in order to maintain the stasis of the universe as it was wrongly related to him by contemporary astronomers [1]. As it turns out, the cosmological constant ultimately relates to the presence of dark energy but not in any way remotely resembling what Einstein would have anticipated.

Fortunately, before Einstein went too far with the cosmological constant idea, an American astronomer by the name of Edwin Hubble (1889–1953) came along and showed that, in fact, the universe was expanding!

Expansion means that the distance between any two points in the universe increases with time, but they do so in relation to a single time-dependent scale factor that we shall denote $a=a(t)$. It is crucial to note that the scale factor is the same in all of space and only varies with time. In rigorous terms, the expansion idea may be formalized as follows: Let $d_{ij}(0)$ be the distance between any two points labeled "i" and "j" in the universe at a particular time, conventionally set to be $t=0$, then the distance $d_{ij}(t)$ at a future time t becomes: $d_{ij}(t)=d_{ij}(0).a(t)$, and this equation holds for any pair i, j of points in space. The important thing is that there is only one scaling factor, and it is not dependent on the pair of points chosen. Hubble noticed that this scaling factor increases with time.

As the concept settles down in our minds, let us drop subindices and adopt a more agile notation, denoting $y=y(t)=d_{ij}(t)$ *and* $\Delta y=d_{ij}(0)$. *Then,* $y(t)=\Delta y.a(t)$, and dy/dt, the rate at which distance y changes in time, becomes $dy/dt=\Delta y.(da/dt)$, where da/dt is the rate of change of the scaling factor of the universe. Combining the previous relations we may write:

$$\frac{dy}{dt}=\Delta y\frac{da}{dt}=y\left[\frac{(da/dt)}{a}\right] \tag{3.1}$$

The factor $\left[\frac{(da/dt)}{a}\right] = H$ on the left-hand side is the Hubble constant of the universe, that is, the rate at which the scale factor changes as a result of the universe expansion divided by the scale factor [1]. It is important to point out that H is constant in space but not in time. Equation (3.1) has far-reaching consequences: If we know the velocity dy/dt of a receding galaxy, we can calculate y, that is, how far away it is. Furthermore, as we shall show later in the discussion, *the more distant the galaxy is from us, the faster it is traveling away from us.*

This is the Hubble picture of the expanding universe, and in this picture, space itself is expanding. It has to, implying that the vacuum is getting larger and larger, a troubling runaway scenario by no means well understood.

Let us elaborate on this context further by applying the well-worn Newton's law of gravitation. The gravitational pull exerted on a galaxy revolving at a distance y from an arbitrarily chosen point in the universe would be $F=GMm/y^2$, where M is the total mass of matter contained within the ball of radius y centered at the chosen point, and m is the mass of the revolving galaxy. The potential energy (U) of the galaxy is the physical magnitude whose rate of change in time is the gravitational force, therefore $U=-GMm/y$. Now, the total energy (E) of the galaxy must include the potential energy associated with the gravitational force exerted on it plus another contribution (K) that represents the kinetic energy, that is, the energy associated with the movement of the galaxy. Let us assume the latter form of energy is only associated with the expansion of the universe. Then, we may write $E = K + U = 1/2m\,(dy/dt)^2 - GMm/y$. But E is unchanged in time unless another force starts working on the galaxy, which we assume not to be the case. So, we may say that the quantity $[m(dy/dt)^2 - 2GMm/y]$ is a constant or $[(dy/dt)^2 - 2GM/y]$ is constant since the mass m remains unchanged. Now, plugging in the previously obtained Equation (3.1) into the last expression, we obtain:

$$y^2\left(\frac{da/dt}{a}\right)^2 - \frac{2GM}{y} = \text{constant} \tag{3.2}$$

The mass enclosed by the sphere of radius y is $(4/3)\pi y^3\rho(t)$, where $(4/3)\pi y^3$ is the volume of the ball of radius y, and $\rho(t)$ is the mass density that obviously keeps decreasing in time in an expanding universe. We may then write:

$$y^2\left(\frac{da/dt}{a}\right)^2 - \frac{2G}{y}\frac{4}{3}\pi y^3\rho(t) = \text{constant} \tag{3.3}$$

A little algebra (division by y^2) and substitution with known relations derived above yields:

$$\frac{\left(\frac{da}{dt}\right)^2}{a^2} = H^2 = \frac{8\pi G}{3}\rho(t) - k/a^2 \tag{3.4}$$

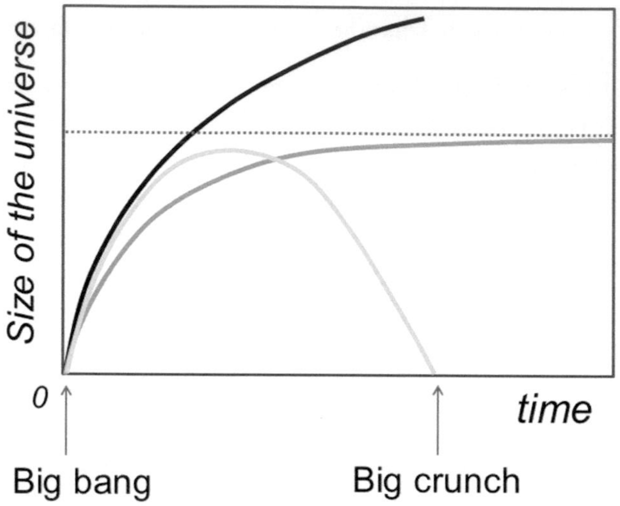

FIGURE 3.1 Qualitative scheme of the time dependence of the size of the universe for a closed universe (light gray), open universe (black), and flat universe (gray).

In this equation, k on the right-hand side is a constant. Equation (3.4) is known as the Friedman-Robertson-Walker (FRW) formula, derived from the theory of general relativity [1]. But we have shown that a simplified version of the formula could be obtained from Newtonian physics, making very simple assumptions.

The reader is once more reminded that H in Equation (3.4) stands for the Hubble constant, which is a constant in space and not in time. Notice that, generally, the following relation holds: $8\pi G/3\,\rho(t) - k/a^2 \geq 0$. This relation becomes the hallmark of an "open universe." On the other hand, if at some point in time we get $8\pi G/3\,\rho(t) < k/a^2$, the universe would stop expanding and start contracting, and we would call it a "closed universe." In this case, the universe would eventually collapse onto itself into a singularity under the gravitational pull. This scenario has been termed the "big crunch." If on the other hand $k=0$, the universe will expand at an ever-decreasing rate because the matter density is monotonically decreasing as the universe keeps expanding. The rate of decelerated expansion will approach the asymptotic limit value of zero at infinite time. This type of universe is called a "flat universe." The time dependence of the size of the three types of universe – open, closed, and flat – is schematically shown in a qualitative fashion in Figure 3.1.

3.2 PHYSICAL PICTURE OF AN EXPANDING UNIVERSE FROM THE BIG BANG UNTIL TODAY

The best up-to-date observations of our skies support the picture of a flat universe ($k=0$); that is, it will continue to expand only until it reaches a particular size. To compute the density ρ, let us assume a cubic box of dimension a in a flat universe containing mass M. Then, the FRW formula becomes

$$\frac{\left(\frac{da}{dt}\right)^2}{a^2} = H^2 = \frac{8\pi G}{3}\frac{M}{a^3} \tag{3.5}$$

This equation can be simplified to $da/dt = \sqrt{8\pi GM/3a}$, which can be rewritten as

$$\left(da/dt\right) = Wa^{-1/2} \tag{3.6}$$

In Equation (3.6), the constant W is defined by parameters of the universe: $W = \sqrt{8\pi GM/3}$. Now, to find out how the universe expands in time, we need to compute the time dependence of $a{=}a(t)$. This may be simply obtained by noting that Equation (3.6) may be rewritten as $a^{1/2}da = Wdt$ giving $a^{3/2} = w't$, or equivalently, we may state that our universe is expanding as

$$a = wt^{2/3} \tag{3.7}$$

where w' *and* w are constants. This is the *predicted* expansion behavior for our current matter-dominated flat universe. As we shall see, the experimental results reveal a huge discrepancy with this rigorously obtained prediction, and dark energy is at the core of the problem!

At the beginning of its existence, right after the Big Bang, the universe would not have been matter-dominated but rather radiation-dominated, with spontaneous creation of matter and its compensatory antimatter that would mutually annihilate with a huge production of photons. A residual amount of matter would have been created as a proxy for the energy released, as described by Einstein's formula $E{=}mc^2$, derived in the previous chapter. In this early universe, let us consider a cube of photons of dimension a. How would this scale factor for the expansion in the early radiation-dominated universe behave in time?

To describe this early universe, we need to resort to some basic principles of quantum mechanics. We have previously discussed that the photon is a massless particle, and yet it carries momentum. This strange duality was something that Einstein found particularly irksome because he could not reconcile that fact with the fact that the photon transfers energy as it hits a surface. He eventually resolved the paradox admirably, as discussed in the previous chapter. Einstein's results in fact imply that the notion of particle needs to be refined. We can no longer think of an elementary particle as a "corpuscle" but rather as a "wave-like excitation" or a "warp in a field." The photon travels at the speed of light, but its "kinetic energy" depends on how many crests are packed in the wave per unit of time; this is what we know as frequency. Max Planck showed that in fact the energy of the photon may be written as $E{=}hf$, where f is the frequency and h is Planck's constant. In fact, h is one of the parameters for our universe, together with G, Newton's gravitational constant, and a few other constants. Together they define how our universe behaves.

Planck's expression for the energy of a photon may be rewritten as $E{=}hc/\lambda$, where c is as usual the speed of light and λ is the wavelength, that is, the distance between two consecutive crest peaks of the "wave particle." The equivalence between both expressions for the energy of the photon follows from the simple fact that, by definition, $f=c/\lambda$. As a increases with the universe's expansion, so does everything else, including the wavelength of the photon, which becomes commensurate with a. The photons in the early universe had their wavelengths stretched as the universe expanded, implying a dramatic cooldown following the Big Bang since the energy goes down as the wavelength increases (the frequency decreases). The wavelength of the "cold" ancient photons is today in the microwave region, a very low-energy region of the radiation spectrum. This is the "glow" we see in the skies known as cosmic microwave background (CMB) radiation, as discussed in Chapter 1. The CMB is the relic of the early radiation-dominated universe that followed right after the Big Bang.

Since the wavelength of the ancient photons can be made proportional to the expansion factor in the radiation-dominated early universe, we may write $E=J/a$, where J is a constant. This gives a radiation energy density $\rho_r = (J/a)(1/a^3) = J/a^4$. The a-dependence in this expression can be contrasted with the previously obtained density for a matter-dominated "later" universe: $\rho = M/a^3$.

We can substitute the expression for the density in the radiation-dominated universe in the FRW formula given by Equation (3.4), keeping in mind that our universe is flat and hence $k=0$. This substitution gives

$$\frac{\left(\frac{da}{dt}\right)^2}{a^2} = \frac{8\pi GJ}{3a^4} \tag{3.8}$$

Equation (3.8) can be rewritten as $ada=Cdt$, where $C = \sqrt{8\pi GJ/3}$, yielding the following time dependence for the expansion in the radiation-dominated early universe: $a \propto t^{1/2}$. This behavior is different from the expansion of the matter-dominated universe, which, as we may recall, has a time dependence of the form $a \propto t^{2/3}$.

The two predicted modes of expansion of the universe can be best visualized in qualitative graphs. Thus, Figure 3.2 shows density versus expansion scale factor a for the radiation-dominated and the matter-dominated universes. There is a cross-over point at approximately 10,000 years from the Big Bang, since the density of the radiation-dominated universe falls more steeply than the density of the matter-dominated universe as a increases. Alternatively, we may plot the size of the universe against time (Figure 3.3). The radiation-dominated universe size increases more slowly (as $t^{1/2}$) than the matter-dominated curve with $a \propto t^{2/3}$. So the latter curve crosses the former at the point determined previously to be in the order of 10,000 years. As previously discussed, the expansion curve for the matter-dominated universe, which is the universe we live in today, will reach an asymptotic limit. This prediction, as we shall see subsequently, is at odds with experimental observations that reveal an accelerated expansion of the universe. This discrepancy between prediction and experiment will be shown to be at the core of the dark energy controversy.

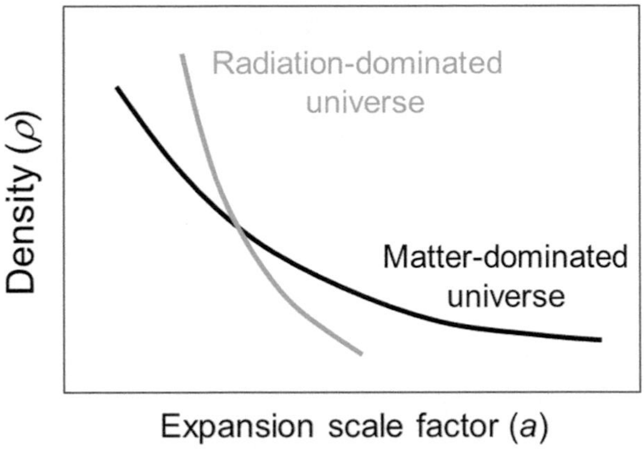

FIGURE 3.2 Qualitative behavior of density in an expanding radiation-dominated and matter-dominated universe.

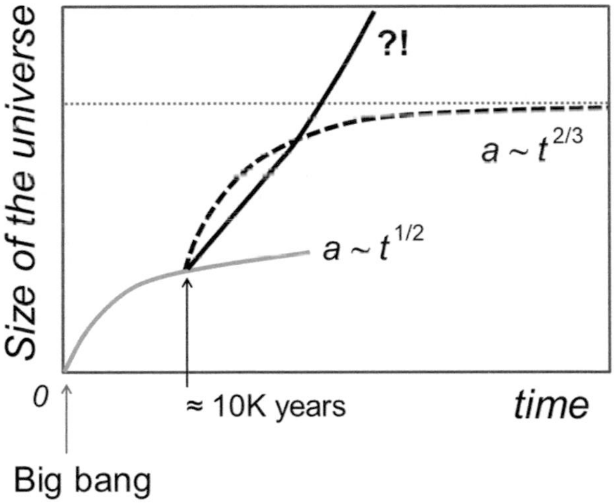

FIGURE 3.3 Qualitative behavior of the time dependence of the size of the universe in a radiation-dominated regime (gray line), matter-dominated regime (dashed line), and experimentally validated runaway regime (dark solid line).

In summary, general relativity (FRW equation) predicts that the universe has expanded under radiation dominance for about 10,000 years and then it expanded faster under matter dominance, and this expansion will slow down to reach an asymptotic limit at infinite time, as described in Figure 3.3.

As we keep anticipating, this is not what seems to be happening according to experimental evidence. The universe's expansion is not showing signs of slowing down. Quite the contrary, the expansion is accelerating, prompting us to invoke the presence of a form of energy of unknown origin that is fueling the runaway. This energy of unknown nature has been ominously termed "dark energy" and may well become the nemesis of our lofty theories on how the universe works [2].

3.3 THE UNIVERSE IN A RUNAWAY MODE

The accelerated expansion of the universe constitutes the biggest challenge to the prevailing paradigm in physics, as it demands either a complete revision of the physical laws to account for this "anomaly," or alternatively – but no less painfully – that we reckon the existence of dark energy, a form of energy of hitherto unknown origin that is constantly being injected into the universe and is causing its runaway behavior in complete defiance of gravitational laws.

The runway picture of the universe was put forth by Perlmutter, Schmidt, and Riess, as discussed in Chapter 1 [1]. They were able to establish the exponential expansion by precisely measuring the so-called redshift effect in the radiation collected from very bright objects called *supernovae*. As the light-emitting source travels away from the observer, the detected frequency f of the emitted light gets lower because the time of arrival of successive wave crests gets longer (Figure 3.4). The net effect can be represented as a stretching of the wave. The observed frequency f is related to the emitted frequency f_e by the relation: $f = f_e\left[\frac{c}{c+v_s}\right]$, where v_s is the velocity of the light source. The apparent stretching of the wavelength is known as redshift since the red color of the visible spectrum corresponds to the longest wavelength and lowest frequency. Thus, the redshift is implied by the relation $f<f_e$.

Conversely, if the light-emitting source approaches the observer at speed v_s, the time of arrival of successive waves is increased, so the waves are bunched together, compressed as

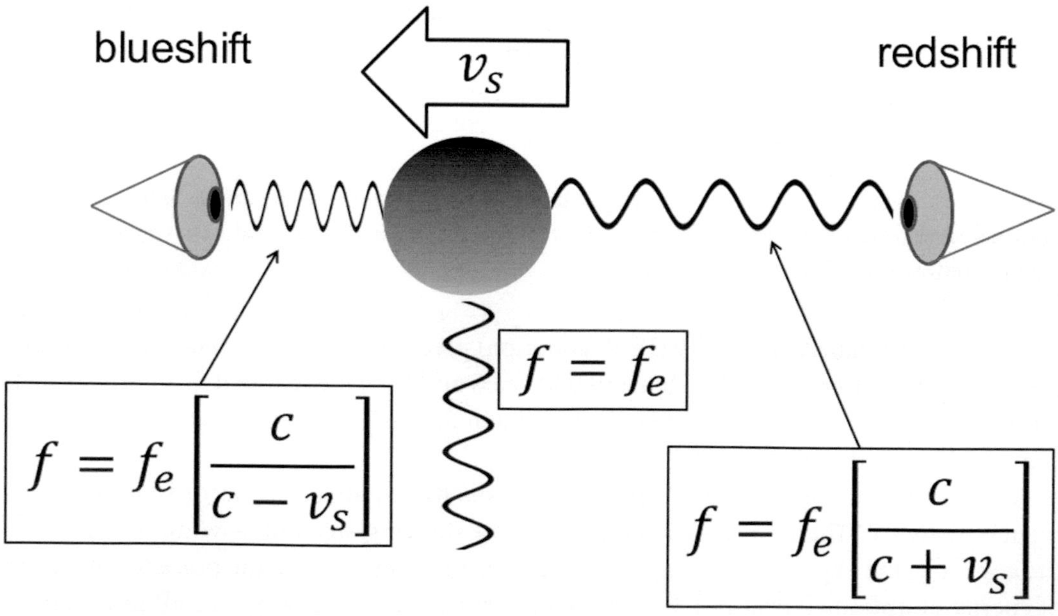

FIGURE 3.4 The redshift and blueshift effect on observed radiation from a light-emitting source moving with velocity v_s relative to the observer.

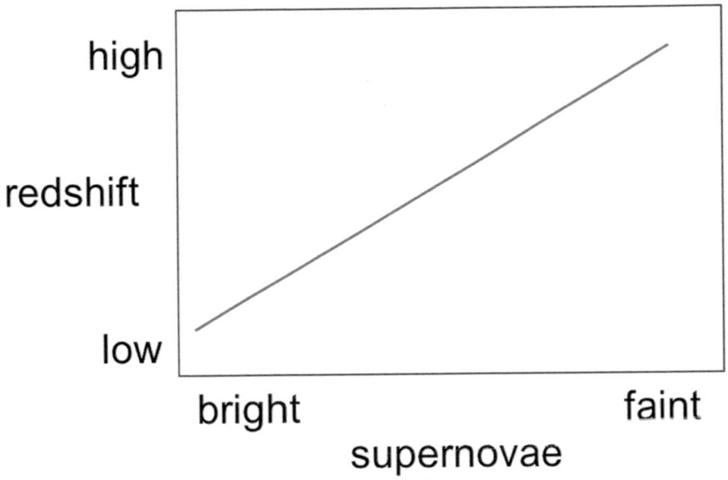

FIGURE 3.5 Qualitative behavior of redshift for nearby and distant light sources. Brightness is accepted to be a proxy for distance. Thus, far-away objects move away from the detector at a faster speed than objects nearby. This behavior constitutes the hallmark of the runaway universe as validated in the Perlmutter-Schmidt-Riess experiments.

it were, with a net "blueshift" effect resulting in increased frequency ($f>f_e$) in accord with the formula $f = f_e\left[\frac{c}{c-v_s}\right]$ (Figure 3.4).

The accelerated expansion of the universe was established by determining the commensurability of the redshift effect with the brightness of the light-emitting source, a proxy for the distance to the observer on Earth. It was determined that the brighter the source, the smaller the redshift, and conversely, the dimmer the source, the more pronounced the redshift (Figure 3.5). The implications of this observation remain as troublesome as they are transparent: The more distant the source, the higher the velocity at which it is traveling away from the observer; in other words: $\nu_s = \nu_s(d) \propto d$, where d is here the distance to the light-emitting source. But this implies that $da/dt \propto a$. The solution of this equation yields an exponential or runaway time dependence: $a \sim e^{qt}$, with q=constant.

This means that the scale factor a is growing exponentially in time, completely at odds with the predicted scenarios of general relativity described previously. We are in the presence of the hallmark for a runaway universe!

3.4 THE UNIVERSE RUNAWAY AS FUELED BY DARK ENERGY

In a manner of digression, we must consider here alternative ways of computing the energy E of the photon. We have dealt with energies, but so far we have not incorporated heat or temperature into our physical picture of the universe. Temperature is the physical quantity conjugated to energy in thermodynamics, the science concerned with the interconversion of heat, energy, and work, whose modern foundations were laid out by the Austrian physicist Ludwig Boltzmann (1844–1906).

It is estimated that the photons in the CMB correspond to radiative emissions of an extremely cold black body, currently at a temperature of approximately 3K (–454.27°F or –270.15°C). On the other hand, the energy of a photon may be related to the temperature of emission through Boltzmann's formula $E=kT$, where k is Bolzmann's constant. Now, given that $E=hf=hc/\lambda\approx hc/a$, we may conclude that

$$T=\left(\frac{hc}{k}\right)\left(\frac{1}{a}\right). \tag{3.9}$$

This implies that the universe is cooling down as it is expanding. We also have an estimation of $T\approx3000$°K for the temperature of the universe at a particular epoch after the Big Bang, namely the ionization epoch, when atoms were stripped of their electron shells, with the atomic nuclei embedded in a plasma of electrons. Thus we may compute the ratio $a_{today}/a_{ion}=3000/3=1000$, where a_{today}, a_{ion} are respectively the expansion scales today and at the ionization epoch. This means that our universe is 1000 times bigger now than in that hot era right after the Big Bang.

Now, how far back did the ionization take place? We can easily answer this question since we know the time dependence of a in a matter-dominated universe and the current age of the universe (t=13.8 billion years, counted since the Big Bang):

$$\frac{a_{today}}{a_{ion}}\approx1000=\left(\frac{t}{t_{ion}}\right)^{\frac{2}{3}}=\left(13.8\times10^{9}\,y/t_{ion}\right)^{2/3} \tag{3.10}$$

This implies that the ionization era occurred at the time $t_{ion}\approx460{,}000$ years after the Big Bang.

Here we reproduce the FRW formula for our flat universe but in a way that accommodates the radiation-dominated regime ($\rho\sim1/a^4$) as well as the matter-dominated regime ($\rho\sim1/a^3$):

$$H^2=\frac{\left(\frac{da}{dt}\right)^2}{a^2}=\frac{8\pi G}{3}\rho(t)=\left(\frac{8\pi G}{3}\right)\left(\frac{C}{a^{3(b+1)}}\right) \tag{3.11}$$

In this equation, C is a constant, and $b=1/3$ for the radiation-dominated regime and $b=0$ for the matter-dominated regime.

As we have anticipated, *these regimes postulated by the theory of general relativity do not agree with the experimentally verified exponential expansion of the universe determined by Perlmutter, Schmidt, and Riess* (see discussion in Chapter 1). These scientists have shown that a behaves exponentially in time: $a\sim e^{\theta t}$ with θ=constant. Now, from Equation (3.10) it follows that the case where $b=-1$ corresponds to a constant density $\rho=C=\rho_0$. *Paradoxically, the $b=-1$ regime not anticipated by general relativity is precisely the type of universe whose* *expansion was measured by Perlmutter, Schmidt, and Riess.* By plugging in b=–1 in Equation (3.11), we get

$$\frac{\left(\frac{da}{dt}\right)^2}{a^2} = \frac{8\pi G}{3}\rho_0 \approx \Lambda$$

$$\Lambda = \text{"cosmological constant of the universe"} \tag{3.12}$$

This formula yields the exponential expansion of the universe, with the fudge factor in Einstein's theory, namely the ill-fated "cosmological constant." We may rewrite it as

$$\frac{da}{dt} = a\sqrt{\Lambda} \tag{3.13}$$

Equation (3.13) is clearly indicative of a runaway process: The larger the expansion, the larger the rate of expansion. As expected, the equation's solution yields a factor a that is exponentially increasing in time (sound familiar?):

$$a \propto e^{\sqrt{\Lambda}t}, \quad \theta = \sqrt{\Lambda} = \sqrt{\frac{8\pi G\rho_0}{3}} \tag{3.14}$$

In other words, by taking b=–1 or fixing the density as constant, we can reproduce the experimental result of exponential expansion and therefore accelerated expansion of the universe that earned Perlmutter, Schmidt, and Riess the Nobel Prize in physics. Furthermore, we can calculate the cosmological constant, bestowing physical meaning to this troublesome factor that Einstein regarded as a fudge term ("my biggest blunder," he called it) in his theory of general relativity.

On the other hand, we know that the speed $v=dy/dt=\Delta y.da/dt$ at which a galaxy runs away from us increases as the distance $\Delta y.a$ to the galaxy increases. But this is precisely what Equation (3.13) is telling us: $v = dy/dt = \Delta y\, da/dt = \Delta y a\sqrt{\Lambda} = y\sqrt{\Lambda}$ or $v \propto y$. Furthermore, the equation is also telling us that the proportionality factor is the root square of the cosmological constant!

One important consequence of this result is that beyond a critical distance $y^* = c/\sqrt{\Lambda}$, the speed of a galaxy at distance $y>y^*$ will be larger than the speed of light: $v>c$! This means that such a galaxy will travel out of sight: Its light will never reach us. The galaxy will simply disappear beyond our horizon. The accelerated expansion of the universe will make galaxies disappear beyond our horizon, hence heralding a much duller view of the skies.

We may have lifted the veil of nature but only to unravel yet another mystery. If ordinary matter (b=0) or ordinary radiation (b=1) are not dominant in the universe, what is causing the accelerated expansion of our universe? A mysterious energy of unknown origin is being constantly created and injected into our universe from a source that does not get diluted as the universe expands. This mysterious energy fueling our universe runaway is what we call dark energy.

At least we know one key thing about dark energy: It does not get diluted as the universe expands since its density is constant ($\rho=\rho_0$). Now, the thing that is constantly being created without getting diluted in the universe as it expands is … vacuum! Indeed, the vacuum does not get geometrically diluted as the universe increases its volume, so dark energy is generated by the vacuum! And we know this is not a negligible contribution: The dark energy density ($\rho=\rho_0$) makes up for about 68.3% of the total density of energy/matter in the universe.

3.5 HOW COULD THE CREATION OF VACUUM TRIGGER THE UNIVERSE RUNAWAY? QUANTUM MECHANICS UNDERPINNINGS OF THE ANOMALOUS BEHAVIOR

Relativity is taking us to unchartered territory far more rarefied even than that of dark matter. The universe expansion is autocatalytic, it has a retro-feeding mechanism for acceleration whereby the more it expands, the more it harvests dark energy, which in turn fuels more expansion, creating a runaway out of vacuum generation.

The universe runaway is the inevitable truth that emerges from the current physics paradigm. The self-stimulated vacuum energy creation becomes a major hurdle in our trend of thought and prompts us to prod the complementary theory to general relativity: Quantum mechanics.

To unravel how the creation of dark energy may come about, we refer to quantum field theory (QFT), where an electron and its antiparticle, the positron, occasionally emerge spontaneously in the vacuum with the emission of a photon. The particles exist for a brief time and then annihilate each other without a trace unless another event elsewhere in the vacuum leads to an interaction with a component of the triad. Energy conservation is violated but only for the extremely ephemeral lifetime of the particle Δt permitted by the so-called *uncertainty principle*, which posits that $\Delta E \times \Delta t \sim \hbar/2, \hbar = h/2\pi$, where ΔE and Δt represent, respectively, *a priori* uncertainty in energy and time. This principle is one of the milestones of quantum mechanics and asserts that uncertainties in conjugated quantities like energy and time, or position and momentum, balance each other, yielding a constant product. Thus, a short uncertainty in time may yield a huge fluctuation in energy, utterly commonplace in QFT. There is no limit placed by the laws of physics on the scale of this energy fluctuation. Nothing prevents it from occurring on a grand scale. The duration is of course subject to the restriction of the uncertainty principle, which merely implies that the universe has zero total energy, which made the fluctuation possible in the first place. Quantum theory implies that the vacuum should be unstable against large fluctuations in the presence of a long-range, negative potential energy term. Gravitation is precisely such a term. This fact encourages us to believe that vacuum creation stimulates further vacuum creation as dark energy adopts the form of vacuum energy fluctuations.

Because vacuum does not undergo any dilution as the universe expands, quantum vacuum energy becomes the prevailing proxy for dark energy, although this answer is far from satisfactory and prompts further investigation, as shown in Chapter 4.

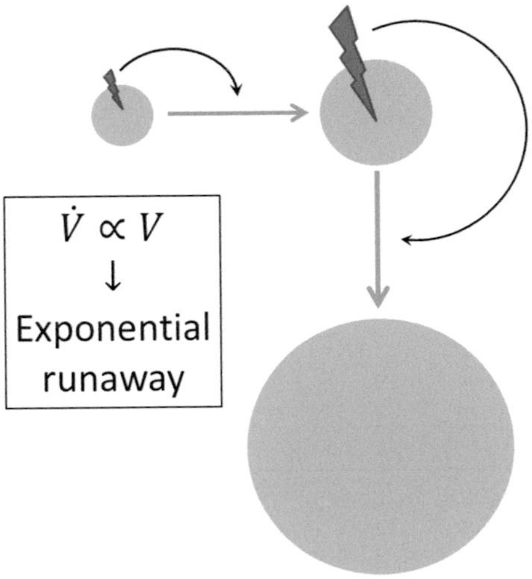

FIGURE 3.6 Kinetic scheme for the principle of autocatalytic vacuum creation (AVC) as the mechanistic underpinning of a runaway universe.

The universe runaway scenario experimentally validated by Perlmutter, Schmidt, and Riess in their Nobel-winning work is at least compatible with the principle of autocatalytic vacuum creation (AVC) hereby put forth (Figure 3.6). The AVC principle can be defined as follows: A quantum vacuum fluctuation fuels the creation of a bigger vacuum, which in turn has an enhanced chance to spontaneously generate a larger quantum fluctuation, which in turn fuels the creation of a bigger vacuum, so the rate of vacuum volume creation $\dot{V} = \dot{V}(t) = dV/dt$ at a given time t is proportional to the vacuum volume $V=V(t)$ created at that time. This assertion may be written as $\dot{V} \propto V$. This equation implies a runaway in vacuum creation, as the solution to the equation yields $V \sim e^{Qt}$, where Q is some constant whose relation to the cosmological constant Λ would need to be established.

For now, we are contented with the assertion that the relativistic Equation (3.12) together with the AVC principle provide the physical underpinnings for the runaway universe as fueled by dark energy. Furthermore, this dark energy is expected to be generated in autocatalytic cycles of vacuum creation that are causative of the universe's runaway behavior.

REFERENCES

1. Weinberg S (2008) *Cosmology*. Oxford University Press, New York
2. Clegg B (2019) *Dark Matter and Dark Energy: The Hidden 95% of the Universe*. Icon Books, London

CHAPTER 4

An AI Quest for Dark Matter Calls for an Extra Dimension

Spacetime tells matter how to move;
matter tells spacetime how to curve.

– JOHN A. WHEELER

SUMMARY

This chapter argues for the existence of a fourth spatial dimension that may account for the storage of dark matter. The quest for dark matter becomes contingent on identifying and validating the topology of the universe as a compact multiply connected space that at present may be locally considered almost flat. Reconciling experimental measurements of the gradient temperature field of cosmic microwave background (CMB) radiation, we conclude that a dormant circular dimension may be consistent with a five-dimensional toroidal space-time with extremely large aspect ratios relative to the dormant coordinate. Unlike the case of Euclidean space, this topology has been preserved since the early days of the universe, when the universe was certainly compact. This topological invariance throughout the time evolution of the universe is in accord with general relativity, whereas a present-day Euclidean universe is not consistent with Einstein's theory in that regard.

4.1 GRAVITY, DARK MATTER, AND EXTRA DIMENSIONS

As discussed in the previous chapters, Newton has been credited for providing the first successful model of gravity in his formidable law of universal gravitation. The predictive power of this law constitutes one of the towering achievements of humanity. Yet, when asked about the nature of gravity, he cautiously responded that he would not frame hypotheses. Humanity had to wait until the early 20th century to get such answers, through the first theory of gravity: Einstein's general relativity [1]. This theory recognizes gravity as an acceleration exerted by a body endowed with mass and regards the fabric of space-time as

DOI: 10.1201/9781003385950-4

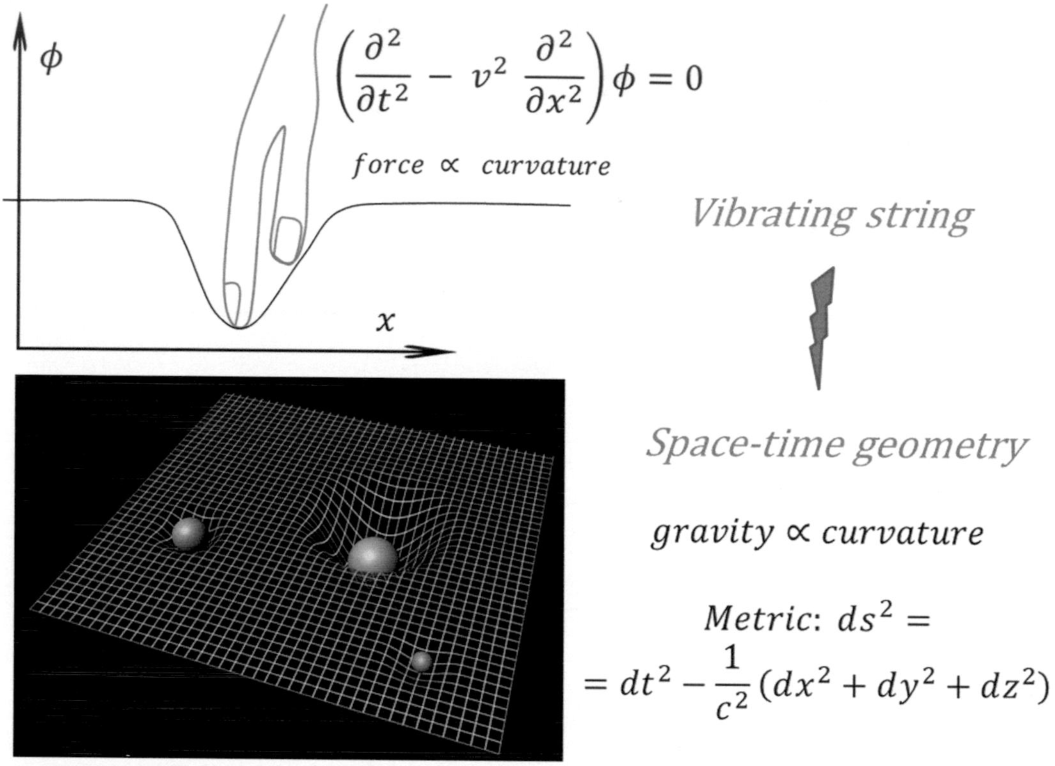

FIGURE 4.1 The equation for the vibrating string where curvature becomes commensurate with the applied force (or acceleration) as the inspirational source for Einstein's differential geometry model of general relativity.

if it were a membrane, whereby acceleration becomes effectively commensurate with curvature, in turn determined by mass concentration (Figure 4.1). Let us examine the equation for a vibrating string that gets perturbed from its equilibrium position along a spatial dimension x by an amount ϕ:

$$\frac{\partial^2 \phi}{\partial t^2} - v^2 \frac{\partial^2 \phi}{\partial x^2} = 0 \tag{4.1}$$

The term on the left indicates the second instantaneous variation of the displacement with respect to time, that is, it represents an acceleration due to an exogenous force (Figure 4.1), while the distortion in the string caused by the force exerted is represented by the curvature $\partial^2\phi/\partial x^2$ of the string. It should be noted that this second spatial derivative is seldom interpreted as the curvature, which in fact it is, just as the first derivative gives the slope of the curve. The proportionality constant is the square of the speed at which the perturbation propagates in time along the x-dimension.

While this fact is typically neglected, it is fairly obvious that the fundamental Equation (4.1) linking curvature and acceleration served as the primary source of inspiration for the differential geometry model of space-time that Einstein adopted in general relativity, his classical theory of gravity (Figure 4.1). It suffices to notice that the metric he adopted in his 4D space-time with coordinates t, x, y, z became:

$$ds^2 = dt^2 - \frac{1}{c^2}\left(dx^2 + dy^2 + dz^2\right) \tag{4.2}$$

This differential volume ds^2 of space-time is the one dictated by the relation between curvature and acceleration, except that the velocity v is now the speed of light c. A complementary way to assert the pivotal relation between curvature and gravity at the heart of Einstein's argument requires that we remind ourselves of another classical equation of physics, the Poisson equation. This equation relates the mass density ρ (amount of mass per unit volume) and the gravitational field ϕ "generated" by the mass. Poisson's equation reads:

$$\nabla^2\phi = 4\pi G\rho \tag{4.3}$$

Where $\nabla^2 = (\partial^2/\partial x^2 + \partial^2/\partial y^2 + \partial^2/\partial z^2)$ and $G=6.674\times10^{-11}\text{Nm}^2/\text{kg}^2$ is Newton's gravitational constant [1]. Thus, integrating Equation (4.3) on a ball of radius r yields Newton's gravitational field

$$\phi = -MG/r, \tag{4.4}$$

where M is the mass contained in the ball of radius r. The dimensions on the right-hand side of Equation (4.3) are those of acceleration (m/s^2), and hence, we again can justify the relationship between curvature (measured as $\nabla^2\phi$) and acceleration due to the gravitational pull per unit volume, given by $4\pi G\rho$. Thus, the differential geometry of the space-time manifold of general relativity is inspired and amply justified by the classical equations of physics. This observation prompts us to assert that Einstein's theory of gravity is in fact a classical theory [1]. It is clear that we now have a theory of gravity that includes its interaction with light, a feature absent from the Newtonian law of universal gravitation. Einstein provides a differential geometry framework in which light travels along a hypermembrane (a manifold often referred to as "brane") whose geometric fabric is defined by mass distribution.

Yet it has proven daunting to reconcile Einstein's theory with quantum physics, now known to govern the other three fundamental forces of nature (electromagnetism and the weak and strong nuclear force). Furthermore, gravity is 10^{-38} times weaker than electromagnetism, an extremely irksome fact that points to a massive geometric dilution of gravity on the known dimensions of space-time. These conundrums are of course compounded by the mysterious nature of dark matter that postulates the existence of invisible massive particles that do not detectably interact with the known elementary particles identified in the "Standard Model" [2], and yet they provide the "missing gravity" in the detectable portion of the universe [3].

A window of opportunity for further scientific inquiry into these seemingly related problems and others arising thereof is offered by the possibility of incorporating an extra (fifth) dimension to the differential geometry of the space-time manifold. To enable the possibility of storing significant amounts of kinetic energy in an undetectable stationary wave amenable to a quantum mechanical treatment, the extra dimension would be expected to be compact, specifically rolled up in a circle of extremely small radius. It should be noted that the smallest conceivable material dimension at present is that of a quark (see next section) of the order of 10^{-18} m (a length unit named attometer) [4].

If we specifically adopt a circular fifth dimension of radius r_0=0.802×10^{-18} m, we find out, through Einstein's relation $E=hf=hc/\lambda$, that it can store stationary waves with extremely large energies

$$E_{5,n} = \frac{n\hbar c}{r_0}, \quad n\lambda = 2\pi r_0, \quad \hbar = \frac{h}{2\pi}, \quad n = 1,2,\ldots \tag{4.5}$$

Strikingly, the lowest such energy is $E_{5,1}$=246GeV, which is precisely the vacuum expectation energy of the elementary particle responsible for bestowing mass on the other known particles, the so-called Higgs boson (see next section) [2]. The daunting problem of incorporating the extra dimension arises from the fact that the circular dimension cannot be thought of as independent of the others, at least there does not seem to be any obvious reason for such an assumption. Hence, the preexisting four dimensions should be considered locally cylindrical or, rather, helical instead of linear, with symmetries becoming only approximate and vast differences in curvature depending on the stride of the helices. Such a universe will be described subsequently and will be shown to be far more suitable to achieve a unified field theory while explaining the extreme geometric dilution of gravity on the "observable space-time manifold," as well as the origin of dark matter and dark energy.

Such a fifth dimension would be extremely hard to detect with current experimental resources, even at the LHC operating at optimal performance, due to the extremely high energies associated with wavelengths of the order of the quark dimension. Yet highly massive particles may be yielded by storing energy on the fifth dimension. To visualize this, we note that the kinetic energy along the fifth dimension would be undetectable and hence will be regarded as rest mass, in accord with Einstein's relation [2]. Thus, the total rest mass M of a particle on a 5D space-time manifold would be given by the equation

$$M^2 = m^2 + m_5^{\,2} = m^2 + \left(\frac{n\hbar}{cr_0}\right)^2, n = 1,2,\ldots \tag{4.6}$$

Here m denotes the detectable rest mass that we would be capable of measuring from the particle's existence in Einstein's 4D space-time, and $m_5=p_5/c$ is the rest mass associated with the component p_5 of momentum along the fifth dimension.

If a particle only stores stationary kinetic energy in the fifth dimension and has no observable rest mass, it is likely to be invisible, yet it could be more massive than any of the known particles. For instance, an "ur-Higgs boson" would have the mass corresponding to the vacuum expectation energy of the Higgs boson, at

$$M = M_{u-H} = \frac{\hbar}{cr_0} = \frac{246GeV}{c^2} \tag{4.7}$$

This is almost twice the detected mass of the Higgs boson, known to be 124.97GeV/c^2. The ur-Higgs and its relatives with masses $M_n = n\hbar/cr_0, n = 2,3,\ldots$ may be regarded as the first particles created after the Big Bang. We may estimate the lifetime at which such particles were created from Heisenberg's uncertainty relation discussed in Chapter 3:

$$\Delta E \times \Delta t \sim \frac{\hbar}{2} \tag{4.8}$$

Since $\hbar = 6.582 \times 10^{-16} eVs$, we may estimate the time of formation of an ur-Higgs boson ($\Delta E = 246 \times 10^9 eV$) at $t = t_{u-H} \approx 1.34 \times 10^{-27}$ s. More massive relatives of the ur-Higgs are expected to emerge at shorter times estimated at $t = t_{u-H}/n$.

As we discuss massive relatives of the ur-Higgs we are prompted to ask: How large can n be? The answer is straightforward, as the so-called Planck mass, $M_P = \sqrt{\hbar c/G} = 1.22 \times 10^{19}$ GeV/c^2, sets an upper bound to the mass, so that if $M_n = n\hbar/cr_0 \geq M_P$, then the particle becomes a black hole. Thus, the first particle formed in the universe had a mass M_{n^*} corresponding to the largest integer n=n* such that $n^* \leq M_P/M_{u-H} \approx 4.95 \times 10^{16}$. This supermassive particle formed at a time estimated at $t^* = t_{u-H}/n^* \sim 10^{-43}s$, during the so-called "Planck epoch" of the universe.

Thus, a picture of the very early universe for the period $10^{-43}s \leq t \leq 10^{-27}s$ may be obtained by assessing the energy stored in the stationary de Broglie wave that spans the fifth dimension, as schematically depicted in Figure 4.2.

4.2 ELEMENTARY PARTICLES AS WARPS IN QUANTUM FIELDS

We now know that atoms are not the indivisible constituents of matter that Democritus and other pre-Socratic philosophers imagined. Atoms are made of even smaller bits, called elementary particles. Yet, popular accounts notwithstanding, there is nothing "corpuscular" about elementary particles when regarded in the current sense of the word. Major modeling efforts that followed after the advent of general relativity and quantum mechanics revealed that the true nature of elementary particles cannot be properly captured in the corpuscular representation. Rather, elementary particles constitute excitations or warps in fields – one for each particle type – that establish a correspondence between each point of space with a scalar or vector value [2]. This is the spirit of the quantum field theory (QFT) that spearheaded current modeling efforts in elementary particle physics.

We are familiar with such vector or scalar fields in physics. For example, a temperature (T) field is a scalar field that assigns a T-value to each point in space. A gradient in T (if the field is smooth or differentiable) steers heat convection, that is, energy transference from the region with higher heat content (at higher T) to the region of lower heat content (at lower T). Thus, we may say that energy (E) is the magnitude associated with the T-field, as reflected in Boltzmann's equation $E=kT$, where k is Boltzmann's constant.

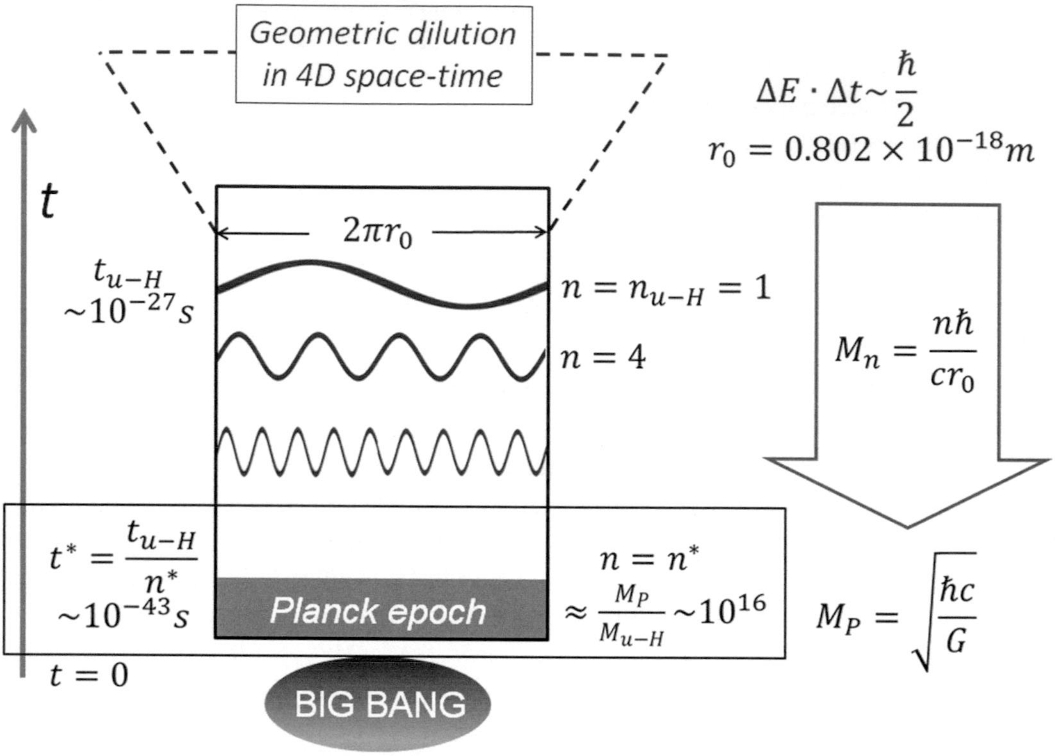

FIGURE 4.2 Elementary particles in an early universe covering the period $10^{-43}s \leq t \leq 10^{-27}s$ after the Big Bang, obtained by storing energy as stationary waves that span the fifth dormant dimension.

Likewise, a dynamical system may be defined by a velocity field, a vector field that determines the trajectories as curves that are tangent at each point to the vector assigned by the field. Similarly to the *E–T* duality, elementary particles have been defined in QFT vis-à-vis fields that represent them. An elementary particle becomes a local concentration of energy or its equivalent mass, as dictated by Einstein's relation $E=mc^2$, as discussed in Chapter 2. The particle field itself is viewed as a map $\phi: W \rightarrow V$, where W is the four-dimensional space-time of special relativity and V is either $\mathbb{R}$, $\mathbb{C}$, the sets of real or complex numbers, or $\mathbb{R}^n(\mathbb{C}^n), n > 1$, depending on whether ϕ is a scalar or a vector field, respectively. Just like E may be viewed as an excitation of its dual T-field, the picture where particles are regarded as excitations of their respective fields is a major tenet of QFT, and the modeling effort resulting thereof is known as the Standard Model (SM) of particle physics [2].

In accord with current thinking, particles are best characterized as local modes of storing energy/mass, that is, localized field excitations defined by independent representations of the symmetry group of space-time [2]. These representations have generators that are parametrized by specific attributes of particles, including charge, mass, spin (an intrinsic angular momentum vis-à-vis an internal axis of rotation), spin degrees of freedom (projection of the spin vector along the direction of particle motion), and so on. Yet, this manner of particle classification tells us very little unless we can define interactions between particles based on their intrinsic attributes, assessing how particles communicate and transform into one another. The nature of these interactions is delineated in the next section.

Now let us briefly explore how the concept of particle field came about. By attributing the energy of the photon to its frequency of oscillation as in the equation $E=hf=hc/\lambda$, Einstein was the first to associate a particle with a wave in his description of the photoelectric effect. Einstein, and Planck before him, reasoned that the frequency was simply a surrogate for kinetic energy, as it is an indicator of the number of cycles per unit time. This was followed by de Broglie's representation of the electron orbiting the atomic nucleus as a stationary wave with a wavelength that satisfies the relation $n\lambda=2\pi r$, where n is a positive integer value and r is the orbit radius. This "quantization" of the wavelength ensures constructive (reinforcing) phase interference, as depicted in Figure 4.3. It also captures the discreteness of the experimentally obtained electron absorption spectrum, as the only allowed transitions for the electron are the result of the absorption of photons carrying energies given by the differences between the n-indexed electron energies $E_n = n\hbar c / r$, with $\hbar = h / 2\pi$. In fact, de Broglie went one step beyond, as he associated any particle with momentum p to a wavelength λ according to the relation $p=h/\lambda$. We can see in these early modeling efforts the need to go beyond a mere corpuscular description of particles, a conceptual framework later adopted by QFT in its construction of the SM [2].

The SM has defined, characterized, and organized all of the known elementary particles in much the same way that the periodic table systematically categorized or classified the types of atoms known as elements. The predictive value of the SM has been staggering, notwithstanding the fact that it does not fully incorporate gravity while being able to encompass the other three quantum forces of nature in a unified way.

At the turn of the 20th century, it was believed that there were only three fundamental particles in nature: Protons, neutrons, and electrons, with the first two making up the atomic nucleus, while the electrons revolved around it, as in a mini planetary system. But

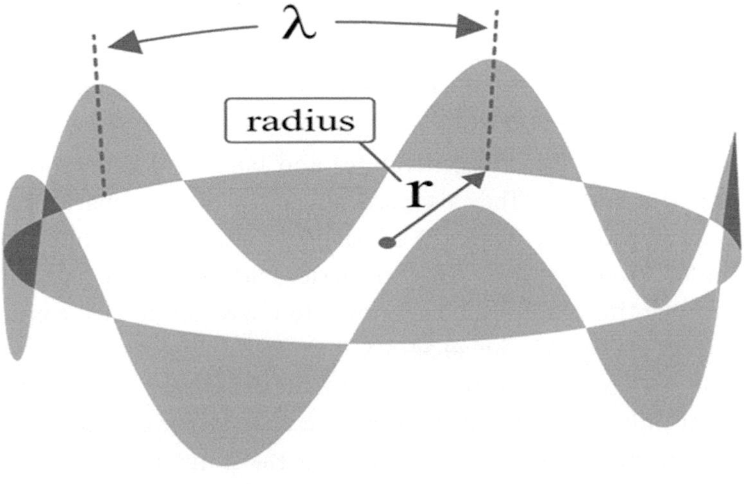

FIGURE 4.3 A de Broglie stationary wave of wavelength λ in a circle of radius r.

by the 1960s, physicists started smashing these particles together and were able to show unambiguously that protons and neutrons were made up of even smaller particles. Many particles were discovered in a relatively short period of time, and for a while, it was not possible to make much sense of the "particle zoo."

Yet, by 1970, mainly due to the monumental "bottom-up" modeling efforts based on QFT, elementary particles were shown to fall into two main categories: Fermions and bosons, with fermions making up matter, while bosons, or more precisely "gauge bosons," transmit forces to which fermions respond or react [2]. Fermions are divided into two kinds of particles, depending on the forces they respond to: Quarks and leptons. Particles within the three types – quarks, leptons, and bosons – are symmetrically related to each other. This means that a coordinate transformation associated with a space-time symmetry operation transforms one particle into another of the same family. It was established that particles communicate with one another via four forces: Electromagnetism, strong nuclear force (SNF), weak nuclear force (WNF), and gravity. The SM describes the first three, while gravity does not yet feature satisfactorily in the current version of the SM, as already anticipated.

Different particles communicate and interact through different forces. For example, only the quarks relate to the gluon, its natural gauge boson, and carrier of the SNF, while electrons relate to the photon, the gauge boson that carries the electromagnetic force. In addition, electrons can also communicate via the W boson and Z boson, the carriers of the weak nuclear force. So electromagnetism is the force that holds electrons in the atom, while the SNF communicated by gluons keeps the nuclei from breaking apart. Meanwhile, the WNF communicated by Z or W bosons assists the radioactive decay of nuclei, whereby, for example, the quark/gluon assemblage of a neutron composite becomes that of a proton upon interaction with a W boson with emission of electromagnetic particles.

Quarks come in six "flavors" that in turn come in pairs to make three generations. These are "up" and "down" (first generation), "charmed" and "strange" (second generation), and "top" and "bottom" (third generation). The up and down quarks are relevant to atomic physics because they make protons and neutrons, while the others make "exotic" matter, too unstable to form atoms and usually lasting a fraction of a second before decaying.

There are six leptons. The best known is the electron, a fundamental particle with a negative charge. The muon (second generation) and tau (third generation) particles are heavier versions of the electron. They also have a negative electric charge, but they are too unstable to feature in ordinary matter. And each of these particles has a corresponding neutrino (Italian for "small neutron"), a spin carrier with no charge and almost no mass. Neutrinos interact via the WNF, which is known to cross-talk to electromagnetism as in the radioactive decay events mentioned above.

Quarks and leptons have twin particles of antimatter known as antiparticles. Antimatter differs from matter only in that it has the opposite charge. For example, the electron has an antimatter counterpart, the positron, which has the same mass but a positive charge. When a particle meets its antiparticle, they both annihilate in a burst of energy carried by a boson.

We shall now characterize the Higgs boson, a particle endowed with a field ϕ that may be represented as a doublet of complex numbers: $\phi=(\phi_1, \phi_2)$, with $\phi_j=Re\phi_j+iIm\phi_j$, $j=1,2$. This field, empirically determined in a brilliant modeling effort, is peculiar in that its potential energy does not reach its minimum at a zero field but a value of ~246GeV (1eV=1.602×10^{-19}J, 1GeV=1.602×10^{-10}J) known as the vacuum expectation value (v), hence implying a symmetry breaking (Figure 4.4). The symmetry breaking arises because different changes in potential energy around the zero-point energy are obtained depending on the direction of variation of the field. For example, if we write the singlet field in radial coordinates as $\phi=\rho e^{i\theta}$, then the zero-point energy is the circle $\rho=\nu$ and radial variations $\rho=\nu\pm\delta h$ yield variations in the potential energy (dashed line in Figure 4.4), while angular variations along the circle $\rho=\nu$ yield no change in potential energy, which remains at its zero value. This symmetry breaking turns out to be of paramount importance in determining the role of the Higgs boson vis-à-vis its interactions with other particles. Thus, just as other gauge bosons are force carriers, communicating different types of forces associated with particle interactions, the Higgs boson may be interpreted as bestowing mass upon other particles.

To interpret this process of mass endowment as well as the communication of fundamental forces by the other gauge bosons, it is necessary to resort to the mathematical arsenal of particle physics [2]. This arsenal is necessary to describe the "geodesic flow," that is, the lines of least action followed by the fields that underlie particle physics. As said, a basic tenet of QFT indicates that a particle is represented as a local excitation of its respective field, representing a local concentration of energy or, equivalently, of mass. In QFT, a particle scalar field $\phi=\phi(\{x_\mu\}_\mu)$ in space-time is determined by the particle Lagrangian $\mathcal{L}=\mathcal{L}(\phi,\{\partial_\mu\phi\}_\mu)$ ($\partial_\mu\equiv\partial/\partial x^\mu$) whose definition is inspired by classical mechanics: $\mathcal{L}=1/2\,(\partial_\mu\phi)^2-V(\phi)$, with the first term on the right-hand side representing the kinetic energy (K), and the second, the potential energy. The latter is exemplified in Figure 4.4 for the Higgs field. Einstein's convention of summation over repeated indices is followed throughout. Thus, the geodesic flow in space-time defined by the field $\phi=\phi(\{x_\mu\}_\mu)$ minimizes the action $\mathfrak{A}=\int\mathcal{L}\,(\phi,\partial_\mu\phi)\,dx^\mu$, and hence the field satisfies the so-called Euler-Lagrange equations:

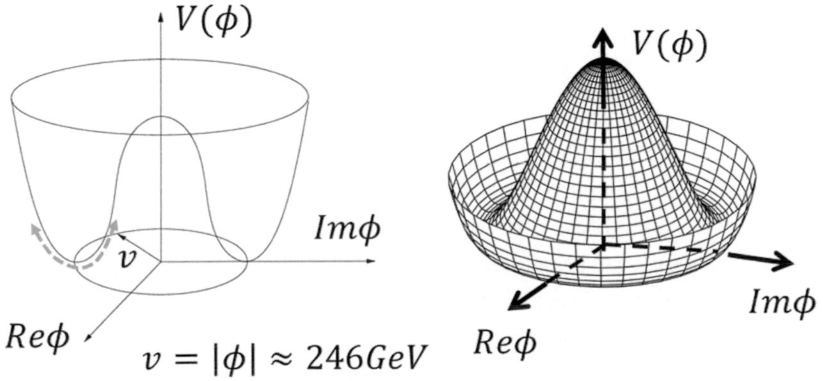

FIGURE 4.4 Scalar field for the Higgs boson.

$$\partial_\phi \mathcal{L} = \partial_\mu \left[\frac{\partial \mathcal{L}}{\partial(\partial_\mu \phi)} \right] \tag{4.9}$$

To see how two particles, determined respectively by fields ϕ_1, ϕ_2, interact, or how one particle communicates a force to another particle, we need to consider the sum $\mathcal{L}_1(\phi_1,\{\partial_\mu\phi_1\}_\mu)+\mathcal{L}_2(\phi_2,\{\partial_\mu\phi_2\}_\mu)$ of their respective Lagrangians. An appropriate coordinate transformation $x^\mu \to x'^\nu$ and suitable transference or swapping of terms yields the sum $\mathcal{L}_1'(\phi_1',\{\partial_\nu\phi_1'\}_\nu)+\mathcal{L}_2'(\phi_2',\{\partial_\nu\phi_2'\}_\nu)$, satisfying the action equality where

$$\int\left[\mathcal{L}_1\left(\phi_1,\{\partial_\mu\phi_1\}_\mu\right)+\mathcal{L}_2\left(\phi_2,\{\partial_\mu\phi_2\}_\mu\right)\right]dx^\mu = \int\left[\mathcal{L}_1'\left(\phi_1',\{\partial_\nu\phi_1'\}_\nu\right)+\mathcal{L}_2'\left(\phi_2',\{\partial_\nu\phi_2'\}_\nu\right)\right]dx'^\nu \tag{4.10}$$

This implies that the original interactive particle pair defined by the fields ϕ_1, ϕ_2, has transformed through mathematical manipulation (including only term reorganization and change of variables) into new particles interpreted as being defined by the fields ϕ_1', ϕ_2'. These destiny fields are associated with the "destiny Lagrangians" $\mathcal{L}_1', \mathcal{L}_2'$ and may correspond to any number of particles, elementary or composite.

For example, let us consider the interaction of the Higgs field $\phi_1=\phi_H$ (Figure 4.4), with a second field ϕ_2 corresponding to a particle that may be the original massless version of the W or Z boson. The kinetic energy K_H for the Higgs field may be written in terms of the covariant coordinates $D_\mu=(\partial_\mu-i\phi_2)$ that respond to the field of the other particle. Upon the coordinate changes $\phi_H\to(\rho,\theta)$ and substituting $\rho=\nu+\xi$ around the ν-value ring of ϕ_H (Figure 4.4), the kinetic energy K_H reads (the asterisk here denotes complex conjugate):

$$K_H = \frac{1}{2}\left(D_\mu\phi_H\right)\left(D_\mu\phi_H\right)^* = \frac{1}{2}\left(\partial_\mu\xi\right)^2 + \frac{1}{2}\phi_2^{\,2}\left(\nu+\xi\right)^2 \tag{4.11}$$

This implies that the precursor to the gauge boson has now gained the potential energy term estimated as $1/2\,\phi_2^{\,2}\nu^2 : \mathcal{L}_2' \approx \mathcal{L}_2 + 1/2\,\phi_2^{\,2}\nu^2$, which implies that the mass bestowed by the Higgs field corresponds to radial (i.e., along the ξ-coordinate) excitations of its field, as depicted in Figure 4.4.

4.3 ALLOWED TOPOLOGY OF THE UNIVERSE ADMITTING A CURLED-UP EXTRA DIMENSION

Einstein's theory of general relativity is formulated in terms of differential geometry, and its laws are therefore cast as differential equations. It is therefore a local theory. It tells us nothing about the overall shape or topology of the universe while it informs on its geometry [5]. In principle, more than one topology can fit the postulated geometric flatness, which at any rate, can only be regarded as an approximation [1, 5]. There is one thing that general relativity does tell us in regard to topology: *As the universe has been evolving in time since the Big Bang, its topology cannot change* [1]. This is fairly obvious since a topological change would imply cutting and pasting space-time, an inadmissible operation. This is so because it would entail a non-homeomorphic distortion of the differential geometry fabric unless a new force beyond the four established forces [2] could be invoked.

In general relativity, we assume that the intrinsic geometry of the universe's 4D manifold has an infinite curvature radius (flat geometry). However, if we soften the assertion to indicate an "immeasurably large radius," we may find alternative topologies for compact (yet enormous) manifolds that would fit the postulated approximate flatness. From the previous discussion, it seemed that identifying dark matter and dark energy would require reverse engineering of the SM to incorporate all dimensions in an early universe together with the evolution of their aspect ratios. Yet, this reverse engineering depends pivotally on the topology of the universe [6], which we know must be invariant in time. This implies that the present-day universe cannot be infinite for the simple reason that the Big Bang singularity and the early universe were compact. *Hence, our universe must be compact, however enormous.*

A quasi-flat compact manifold must necessarily be multiply connected, and hence the incorporation of a fifth circular coordinate yields one and only one manifold as the only alternative: A five-dimensional torus, that is, a Cartesian product of five circles with a huge aspect ratio between the four circular coordinates conforming the locally quasi-flat four-dimensional space-time and the extra fifth spatial coordinate with a quark-size (i.e., LHC-undetectable) radius. We regard the standard four-dimensional space-time in general relativity as a 4-torus locally homeomorphic to the Euclidean space. This is a latent compact manifold representing a quotient space denoted W/~ [6], so that two points in a higher dimensional space W, the five-dimensional torus with a circular extra dimension, would be equivalent to modulo "~" if they projected onto the same point in the latent manifold. The task ahead then becomes to elucidate how dark matter would fit into this scheme as a gravity-carrying particle enshrined in the compact fifth dimension, existing in W but not in its quotient space W/~. In this way, dark matter would not interact with the SM in 4D space-time, except via gravity.

The postulated toroidal topology of the universe has found recent experimental validation in the examination of the cosmic microwave background (CMB), the cosmic relic of the Big Bang (Figure 1.3). If space were infinite (hence flat and simply connected) perturbations in the temperature gradient field of the CMB radiation would exist on all scales. If space is finite, then there would be wavelengths missing that are larger than the size of space itself. Maps of the CMB perturbation spectrum made with probes like NASA's WMAP and the Planck probe from the European Space Agency have shown striking amounts of missing perturbations at large scales. The properties of the observed fluctuations of the CMB show a "missing power" on scales beyond the size of the universe [7]. That would imply that our universe is multiply connected and finite. The spectrum of the CMB is compatible with a three-dimensional torus topology for the spatial components of the universe. Thus, universe models with spatially multiply connected topology contain a discrete spectrum of the Laplacian with a specific wavelength cut-off, as observed in the CMB. Furthermore, the three-dimensional torus model possesses a two-point correlation function that fits CMB maps obtained with the Planck probe [7].

Thus, the toroidal topology enables the incorporation of the extra curl-up dimension as an undeveloped feature in the compact universe evolution. In principle, nothing precludes the presence of more than one such dormant dimension. The dormant dimension cannot be probed with current means, as it spans the quark attometer scale, storing energy in the order of 246 GeV. On the other hand, the flat simply connected universe that we all intuitively have come to grips with would actually be incompatible with the evolution from a compact manifold as implied by the Big Bang scenario simply because such evolution would entail a change in topology. Such a transition from compact multiply connected to flat simply connected cannot be smooth and would be forbidden by general relativity.

Within the compact multiply connected model of today's universe, four of the circular dimensions have grown so much that the curvature radius is for all practical purposes infinite, and the geometry is locally that of a Euclidean space [1]. Thus, the dormant fifth coordinate combines with the quasi-Euclidean standard coordinates in a local cylinder of quark-size radius, where the standard coordinate constitutes the cylinder axis. In essence, the torus with a huge aspect ratio becomes locally a cylinder (Figure 4.5). Standard space-time coordinates get mixed with the dormant coordinate as determined by a pitch angle α defining the helix stride, so that a geometric dilution (υ) for the energy stored in the dormant coordinate may be defined as $\upsilon=-\log \cos\alpha$. Thus, the geometric dilution becomes infinite when there is no projection along the dormant coordinate, and zero when there is no projection onto the standard coordinates.

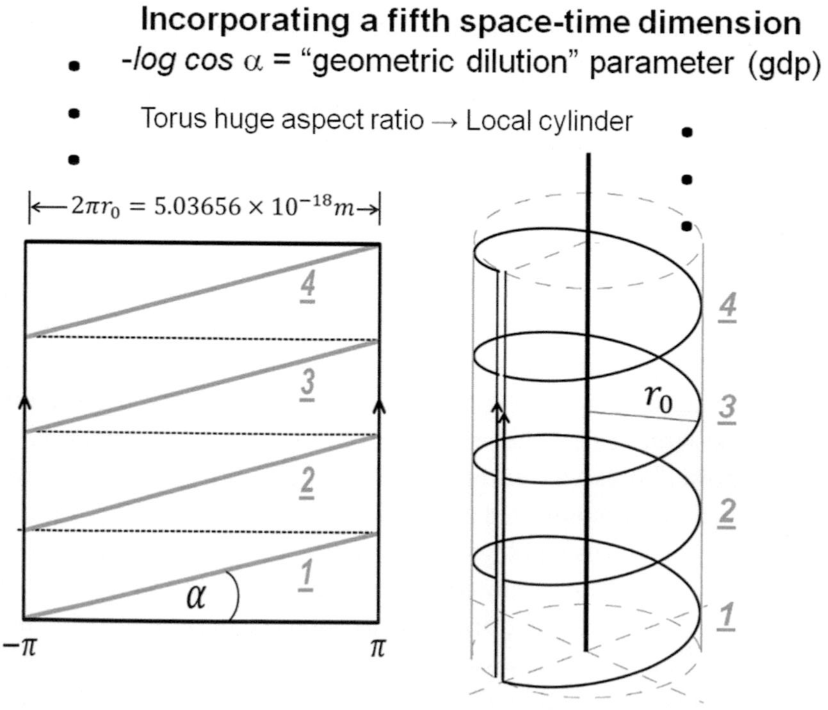

FIGURE 4.5 Incorporating the circular fifth dormant dimension and combining it with generic locally flat dimensions of the four-dimensional space-time through a geometric dilution parameter determining the helical stride.

4.4 A TOPOLOGICAL METAMODEL OF THE UNIVERSE IN THE AI QUEST FOR DARK MATTER

Current knowledge of dark matter is sketchy at best. What we know or can judiciously conjecture is best summarized as follows:

- Dark matter is invisible, massive, "cold" (speed far lower than the speed of light, v<<c), mainly formed early on, interacts with the SM only via gravity, and does not decay easily.
- It is probably not an extension of the SM, as no particle derived from the latter fits the characteristics of dark matter [3].
- It may be stored in a compact fourth spatial dimension.
- In advocating for storing dark matter in an extra dimension, we note that the quark scale ($q=0.802\times10^{-18}$m) is the smallest – undetectable – material scale known and that a stationary wave along a circular fourth spatial dimension with radius q stores an energy $E = hc/2pq = 246$ GeV, the vacuum expectation value of the Higgs boson.
- We have a validated topological model of the compact universe to incorporate such a dimension and trace the origin of dark matter to the early universe.

The core problem in this state of affairs involves the identification of dark matter as stored in the five-dimensional "quintessential" space W, a five-dimensional torus projecting onto a four-dimensional "latent" manifold, Ω=W/~, which is locally flat, meaning it is locally homeomorphic to a Euclidean space. Thus, Einstein's four-dimensional space-time Ω is regarded as a quotient space for the space that we intend to prove stores dark matter. This problem requires the deployment of AI, as it requires the extension of the SM to incorporate a fifth dimension, so far defined in the latent quotient space Ω=W/~. This task, undertaken in Chapter 6, entails the lifting of each pairwise particle interaction within Ω, represented by the transformation $\mathcal{L}_1(\phi_1,\{\partial_\mu\phi_1\}_\mu)+\mathcal{L}_2(\phi_2,\{\partial_\mu\phi_2\}_\mu)\rightarrow\mathcal{L}_1'(\phi_1',\{\partial_\nu\phi_1'\}_\nu)+\mathcal{L}_2'(\phi_2',\{\partial_\nu\phi_2'\}_\nu)$, to the level of W:

$$\widetilde{\mathcal{L}_1}\left(\widetilde{\phi_1},\left\{\partial_\mu\widetilde{\phi_1}\right\}_{\mu=1,\ldots 5}\right)+\widetilde{\mathcal{L}_2}\left(\widetilde{\phi_2},\left\{\partial_\mu\widetilde{\phi_2}\right\}_{\mu=1,\ldots 5}\right)\rightarrow\widetilde{\mathcal{L}_1'}\left(\widetilde{\phi_1'},\left\{\partial_\nu\widetilde{\phi_1'}\right\}_{\nu=1,\ldots,5}\right)+\widetilde{\mathcal{L}_2'}\left(\widetilde{\phi_2'},\left\{\partial_\nu\widetilde{\phi_2'}\right\}_{\nu=1,\ldots 5}\right) \tag{4.12}$$

where $\widetilde{\mathcal{L}_n},\widetilde{\phi_n}$ denote respectively the lifting of Lagrangian $\mathcal{L}_n$ and particle field ϕ_n, with n indicating particle types in the SM. As indicated above, the primes denote destiny particles arising from the interaction.

A lifting of the SM is valid if and only if the diagram in Figure 4.6 is commutative. This property indicates that a canonical projection $\pi:W\rightarrow\Omega$ followed by interaction within the latent space Ω yields a destiny state that is the same as that obtained when the interaction is computed directly in the quintessential space W and the resulting destiny state is

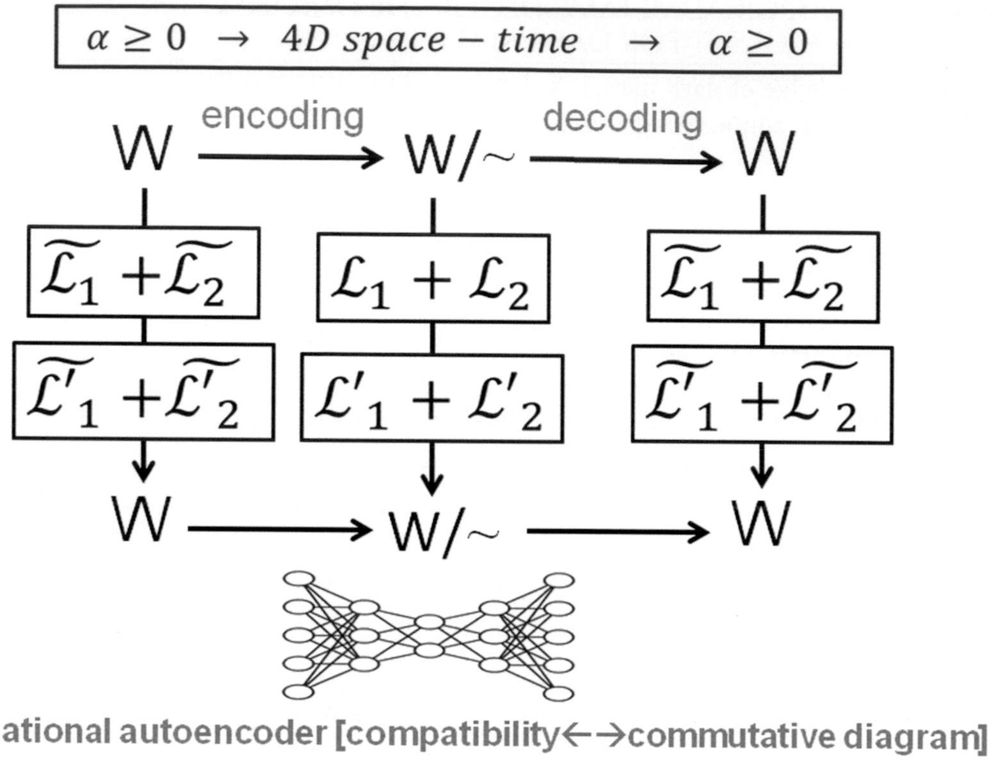

$$x^{\mu} \rightarrow x'^{\nu} \Rightarrow \int [\mathcal{L}_1+\mathcal{L}_2]dx^{\mu} = \int [\mathcal{L}'_1 + \mathcal{L}'_2]\, dx'^{\nu}$$

FIGURE 4.6 Neural network with autoencoder architecture designed to reverse engineer the Standard Model by incorporating a fifth circular dimension.

subsequently projected onto the latent space Ω. In other words, the following equation must hold for every particle pair $(X_J(\mathcal{L}_J, \phi_J), J = 1,2)$ in the SM:

$$X'_j\left(\mathcal{L}'_j, \phi'_j\right) = \pi \widetilde{X}'_j\left(\widetilde{\mathcal{L}}'_j, \widetilde{\phi}'_j\right) \tag{4.13}$$

The lifting $\tilde{\phi}: W \rightarrow \mathbb{C}^n$ of a particle field $\phi: W/\sim \rightarrow \mathbb{C}^n$ requires a particular type of AI system named *variational autoencoder* [6]. Usually, in complex dynamical systems, such autoencoders are used to extract the latent space and obtain the simplified differential equations defined on the latent space as entraining the full system. In other words, the autoencoder is used to simplify the dynamical system, retaining the dynamics that are essential and averaging out subordinated degrees of freedom. Chapter 5 will be devoted to describing such autoencoders, with a special focus on systems that yield topological metamodels.

In the context of extending the SM, or rather reverse engineering it to encompass a dormant dimension, we need to have the autoencoder working in reverse. Rather than simplifying the fields defined in latent space, we need to lift them to a quintessential space of higher complexity, so as to recover the original fields when taking the quotient modulo the standard four-dimensional space-time coordinates. In addition, we need the autoencoder to produce the topological metamodel that would enable the proper lifting, so that a diagram of the type presented in Figure 4.6 becomes commutative.

An illustration of the kind of tasks performed by the topological autoencoder of the SM is depicted in Figure 4.7. The figure describes the lifting to quintessential space W of a process, known as beta decay, comprising the neutron (n) transformation into a proton (p) by the transformation of one of its constitutive down quarks (d) into an up quark (u), with emission of an electron (e⁻) and an electron antineutrino ($\bar{\nu}_e$). This process requires the

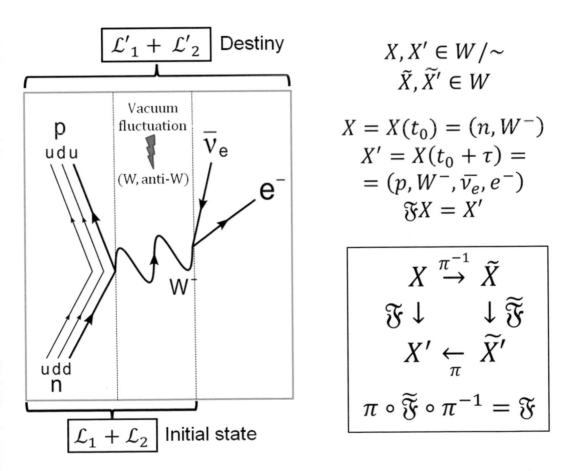

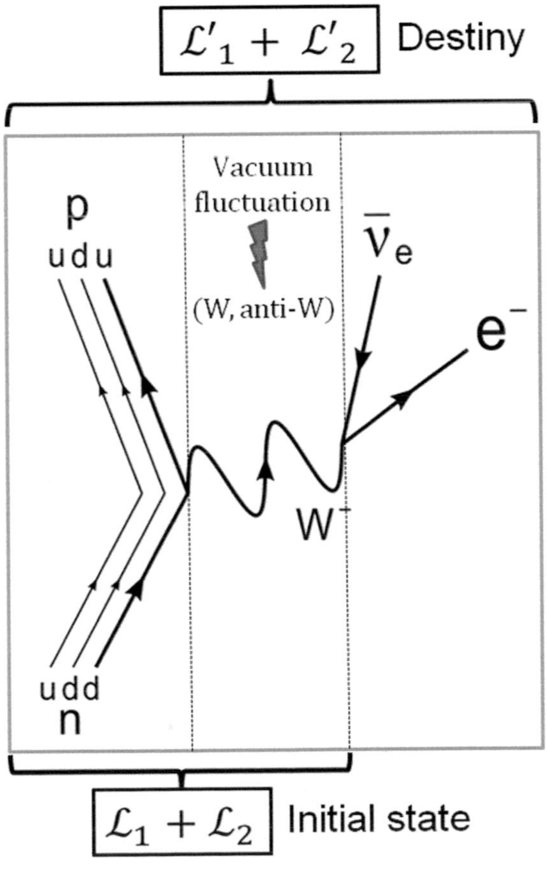

$$X, X' \in W/\sim$$
$$\tilde{X}, \tilde{X}' \in W$$

$$X = X(t_0) = (n, W^-)$$
$$X' = X(t_0 + \tau) =$$
$$= (p, W^-, \bar{\nu}_e, e^-)$$
$$\mathfrak{F}X = X'$$

$$\begin{array}{ccc} X & \overset{\pi^{-1}}{\rightarrow} & \tilde{X} \\ \mathfrak{F}\downarrow & & \downarrow\tilde{\mathfrak{F}} \\ X' & \underset{\pi}{\leftarrow} & \tilde{X}' \end{array}$$

$$\pi \circ \tilde{\mathfrak{F}} \circ \pi^{-1} = \mathfrak{F}$$

FIGURE 4.7 Elementary process of beta decay defined on the latent space representing the four-dimensional space-time W/~ and consistently lifted to the quintessential space W. The process consists of a neutron transformed into a proton through communication with a W⁻-gauge boson with the emission of an electron and an electron antineutrino.

interaction of the neutron with a W⁻-boson, a massive particle (M=80.433GeV/c^2) whose existence is ephemeral ($\tau \approx 10^{-26} s \sim 6.582\times10^{-16} eVs/(2\times 80.433\times10^{9} eV)$), as its formation requires borrowing energy from the vacuum as a fluctuation governed by Heisenberg's uncertainty principle ($\Delta E \times \Delta t \sim \hbar/2$). Here we adopt the notation $X=(n,W^{-}), X'=(p,W^{-},\overline{\nu_e},e^{-})$ to indicate respectively the initial and final state of the system as recorded on the latent space W/~ and dictated by the SM. The respective liftings to W are $\widetilde{X},\widetilde{X'}$, and their validity is ensured if and only if they satisfy the flow relation

$$\left(\pi\circ\tilde{\mathfrak{F}}\right)\tilde{X}=\mathfrak{F}X;\quad \mathfrak{F}X=X';\quad \tilde{\mathfrak{F}}\tilde{X}=\widetilde{X'} \tag{4.14}$$

The operators $\mathfrak{F}$ and $\tilde{\mathfrak{F}}$ determine respectively all processes in W/~ dictated by the SM and the lifting of such processes to the quintessential space W. Thus, the operators may be interpreted as time evolution propagators over time intervals identified with the lifetimes of the particles that communicate fundamental forces. In this context, the autoencoder is charged with the task of computing the operator $\tilde{\mathfrak{F}}$ and canonical projection π, so that the following generic diagram is commutative:

$$\begin{array}{c} W\xrightarrow{\pi}W/\sim \\ \downarrow\tilde{\mathfrak{F}}\quad\downarrow\mathfrak{F} \\ W\xrightarrow{\pi}W/\sim \end{array} \tag{4.15}$$

In this way, the SM may be extended by the topological autoencoder to incorporate an extra dimension compatible with the compact and multiply connected topological metamodel of the universe validated experimentally by the CMB spectrum. The topological autoencoder would then be in a position to discover dark matter particles as it reverse-engineers the SM to adapt it to the early universe, where the dormant dimension becomes commensurate with the other four dimensions of space-time. This program will be pursued in Chapter 6.

REFERENCES

1. Weinberg S (2008) *Cosmology*. Oxford University Press, New York
2. Feynman RP, Weinberg S (1999) *Elementary Particles and the Laws of Physics*. Cambridge University Press, Cambridge, UK
3. Profumo S (2017) *Introduction to Particle Dark Matter*. World Scientific Publishing, Singapore
4. Abramowiczy H, Abtt I, Adamczykh L, Adamusae M, Antonelli S, et al., Zeus Collaboration (2016) Limits on the effective quark radius from inclusive *ep* scattering at HERA. *Phys Lett B* 757: 468–472

5. Hawking SW, Ellis GFR (2023) *The Large Scale Structure of Space-Time: 50th Anniversary Edition*. Cambridge University Press, Cambridge, UK
6. Fernández A (2022) *Topological Dynamics for Metamodel Discovery with Artificial Intelligence*. Chapman & Hall/CRC, Taylor & Francis, London
7. Aurich R, Buchert T, France MJ, Steiner F (2021) The variance of the CMB temperature gradient: A new signature of a multiply connected Universe. *Class Quant Grav* 38: 225005

CHAPTER 5

Methods

Topological Autoencoders for Dynamical Systems in Molecular to Cosmological Applications

The physical world is only made of information,
energy and matter are incidentals.

– JOHN A. WHEELER

SUMMARY

This chapter introduces artificial intelligence (AI) at a fairly elementary level and broadly delineates its possibilities for model discovery, focusing on dynamical systems. The presentation deals mostly with deep learning, autoencoders, and other more specialized architectures, and is tailored to researchers seeking to unravel physical models distilled from big data arising in biological or cosmological applications. With the leveraging of artificial intelligence, dynamical systems have found fertile ground for development. Machine learning identifies parsimonious models providing physical underpinnings of time series data. However, such heavily parametrized models hardly yield physical laws. The problem becomes daunting as we turn to multi-scale biological or cosmological complexities. This chapter addresses these imperatives as it introduces topological methods that enable metamodel discovery and the proper computational tools to decode the metamodel as an inferential framework. The methods advance model discovery, enabling reverse engineering of big data arising in a wide range of cosmological applications, where metamodels with emergent quantum behavior are crucial to providing physical underpinnings of quantum gravity.

The approach is mainly directed at identifying parsimonious metamodels of dynamical systems that describe highly complex contexts. The physical underpinnings of the

DOI: 10.1201/9781003385950-5

processes taking place in such settings do not lend themselves to be described by sparse systems of differential equations, traditionally regarded as the hallmarks of model discovery. The encoding of physical laws within AI-recognizable topological patterns constitutes emerging approaches attuned to effectively process AI-generated patterns within powerful metamodels of significant generality.

This chapter also addresses the problem of quantum gravity as an emergent property in the physics of machine learning. To that effect, the chapter explores the possibility of an AI-based construction of a quantum holographic autoencoder, which requires that we first deal with the physics of machine learning and specifically inquire whether emergent quantum behavior can arise in a neural network. By emergent quantum mechanics, we mean a formulation within a framework of nonlocal equilibrated hidden variables, as in the Bohm scheme. Once an emergent quantum behavior is shown to become possible within the machine learning system equilibrated on the nontrainable – i.e., hidden – variables, we address the question of developing a relativistic string gravitational scheme on the hidden variables adopted. Thus, the network with equilibrated nontrainable variables becomes in effect a quantum gravity autoencoder for the network exhibiting emergent gravity in the non-equilibrium regime prior to the equilibration of the nontrainable variables. In this way, we build a quantum metamodel for gravity that fulfills at least in part a major imperative for physicists seeking a unified field theory. Furthermore, the physical possibility of quantum tunneling across quantum gravity autoencoders supports the idea that our universe may be the progeny of an older universe.

5.1 PRIMER ON NEURAL NETWORKS FOR MODEL DISCOVERY

The leveraging of artificial intelligence (AI) for model discovery in dynamical systems is revolutionizing both disciplines, leading to a mutually beneficial development. The cases of interest in this book are not amenable to model discovery in the sense of yielding a sparse system of differential equations that entrains the full dynamical system. Rather, we are seeking something more elementary and subtle: A metamodel, or topological representation of the dynamics. With the implementation of topological methods, AI-empowered metamodel discovery is able to focus on levels of system complexity and multi-scale hierarchies considered off-limits for current AI technologies. The information on time series is encoded at the maximum level of coarse-graining, hence greatly simplifying the computations while enabling decoding of the information generated at the level of a topological description.

In dealing with dynamical systems using AI-based approaches, we address the following core question: What constitutes an insightful parsimonious model? The standard answer is: "a sparse system of differential equations on latent coordinates." As argued in this chapter, this is not necessarily the format chosen by AI, given the "dimensionality curse" associated with the ultra-complex realities we chose to work on. The deployment of AI requires a paradigm shift, where dynamic information gets encoded in what would be termed a "topological metamodel." The metamodel is essentially pattern-based, where AI-interpretable topological patterns encode physics laws. These methods are likely to advance model discovery as they enable the reverse engineering of time series stemming from vastly complex realities hitherto inaccessible to other AI methods.

AI refers to machines capable of exhibiting behavioral traits that humans regard as indicators of intelligence, such as learning and problem-solving [1]. Within this protean subject, machine learning (ML) refers to the ability to learn without being explicitly instructed to do so, while deep learning (DL) refers to an automated extraction of features, patterns, and ultimately models from arrayed data that is sequentially represented within an abstraction hierarchy organized as a multi-layered neural network (NN) [2–4].

DL has been shown to be highly efficacious at identifying features that are in principle discoverable from the data [2, 5]. As in face recognition, features are hierarchically organized, so that large-scale patterns (eyes, noses, face shapes) emerge after several layers of abstraction from simpler or more rudimentary patterns (lines, curves, shades). The beauty and power of DL reside in the fact that the feature extraction process may be carried out in an unsupervised manner: The features emerge from the training of the system without human input or bias and enable the network to make accurate inferences. In this era of big data, we may state that there are several compelling reasons for implementing DL approaches:

- Fields like biology, particle physics, and cosmology are generating vast amounts of data and time series that can be easily stored and interpreted to achieve a conceptual unification within overarching models.
- We have the right hardware, i.e., graphics processing units (GPUs) that are massively parallelizable.
- We have adequate software such as TensorFlow (TF) that enables suitable modular coding if the data can be pixelized or voxelized into a tensorial array, be it a vector, a matrix, or a tensor proper [3, 4].

At the most basic level, the building block of a NN is the neuron, referred to as *perceptron* [5]. The perceptron enables forward propagation of information encoded in an array of inputs $x_1,x_2,\ldots,x_n$ weighted by parameters $w_1,w_2,\ldots,w_n$ to generate an output of the form $y = f\left(\sum_{i=1}^{n} x_i w_i + w_0\right)$ where w_0 may be regarded as a bias term and f is a nonlinear activation function, often a sigmoid or sigmoid shaped, as shown in Figure 5.1. The bias term enables the activation function to be shifted. In vector representation, we often write: $\boldsymbol{y} = \boldsymbol{f}(z)$, $z = \boldsymbol{x}.\boldsymbol{w} + \boldsymbol{w}_0$, where $\boldsymbol{x}$ and $\boldsymbol{w}$ are respectively input and weight vectors. One can subsequently build a fully connected layer of perceptrons indexed by j, whereby $z_j = \boldsymbol{x}.\boldsymbol{w}_j + w_{0,j}$. The dense layer can be readily implemented in TF code, by simply specifying the number of outputs/perceptrons [3, 4]. We usually refer to the vector of linearly transformed inputs **z** as the "hidden layer," as it does not explicitly describe observables [5].

Hidden layers may be stacked as in DL architectures. Thus, for the j-th perceptron in the k-th hidden layer, we get: $z_j^{(k)} = \sum_{i=1}^{n_{k-1}} y_i^{(k-1)} w_{ij}^{(k)} + w_{oj}^{(k)} = \sum_{i=1}^{n_{k-1}} f(z_i^{(k-1)}) w_{ij}^{(k)} + w_{oj}^{(k)}$. The sequential composition of the network by stacking hidden layers has a standardized script in TF

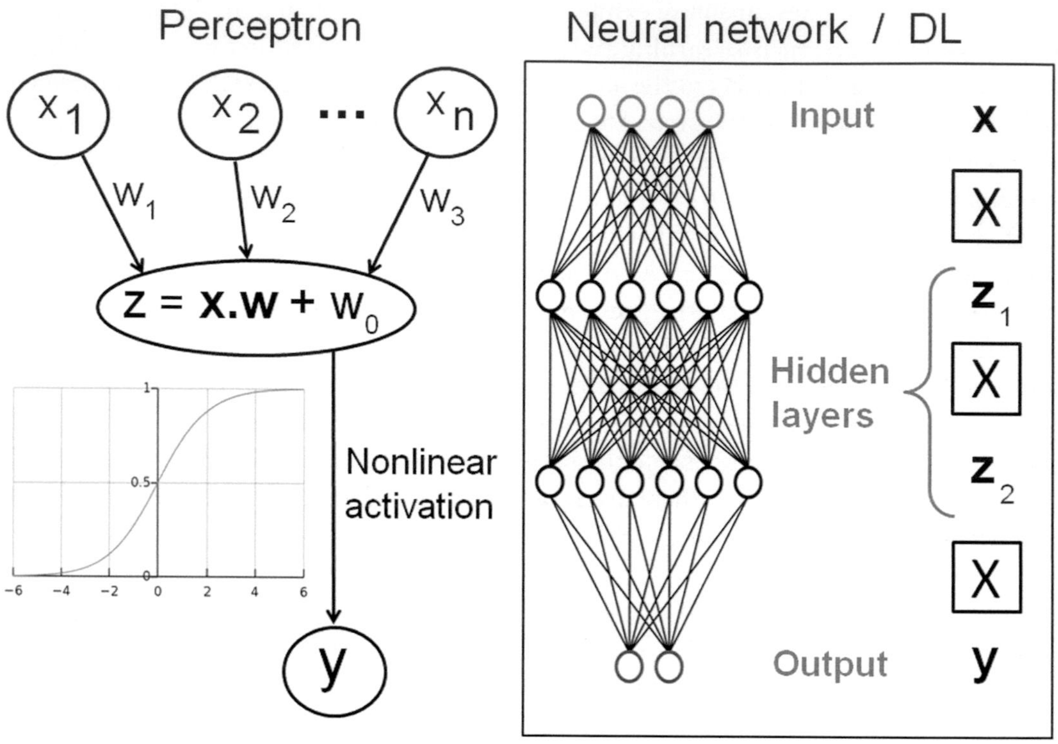

FIGURE 5.1 Scheme of the perceptron or neuron activation by a linear transformation of input stimulus (x) followed by nonlinear signal transmission as output y. The panels on the right show the organization of NNs with one and multiple hidden layers. The "boxed X" indicates full (dense) node connectivity between consecutive layers.

that generates the propagation of information as specified by the equation above [3, 4]. In an NN with K hidden layers, we may regard the output $y_j^{(K)} = f\left(\sum_{i=1}^{n_{K-1}} y_i^{(K-1)} w_{ij}^{(K)} + w_{oj}^{(K)}\right)$ as an "inference" made by the DL system.

The accuracy of the DL inference represents the level of optimization of network performance and may be assessed vis-à-vis a training set of input/output paired data points. This assessment is often referred to as the *loss* of the network [2, 5]. The loss is parametrically dependent on the full weight tensor $\boldsymbol{W} = [w_{i_{k-1}j_k}{}^{(k)}]_{k=1,\ldots,K}$, which may contain a huge number of weights, in the thousands if not millions, depending on the size of the network. Thus, for DL NN with K hidden layers, the loss function or empirical risk $J(\boldsymbol{W})$ becomes $J(\boldsymbol{W}) = |\mathfrak{I}|^{-1} \sum_{\xi \in \mathfrak{I}} \mathcal{L}\left(\boldsymbol{y}_\xi^{(K)}(\boldsymbol{W}), \boldsymbol{y}_\xi\right)$, where $\mathfrak{I}$ is the training set with a number of elements $|\mathfrak{I}|$, $\boldsymbol{y}_\xi^{(K)}(\boldsymbol{W})$ is the predicted output vector for input vector $\boldsymbol{x}_\xi$, $\xi \in \mathfrak{I}$, $\boldsymbol{y}_\xi$ is the actual output vector, and $\mathcal{L}(\boldsymbol{y}_\xi^{(K)}(\boldsymbol{W}), \boldsymbol{y}_\xi)$ measures the discrepancy between actual and predicted output. In regression problems, where the output is a numerical vector, it is often convenient to adopt $\mathcal{L}(\boldsymbol{y}_\xi^{(K)}, \boldsymbol{y}_\xi) = \| \boldsymbol{y}_\xi^{(K)}(\boldsymbol{W}) - \boldsymbol{y}_\xi \|^2$. In such cases, the optimization of the NN through training becomes a problem of least squares. Optimizing the NN is tantamount to minimizing the loss $J(W)$, which requires a careful fine-tuning of the size of the training set vis-à-vis the size of the weight parametrization. In principle, the optimal network is defined as $\boldsymbol{W} = \boldsymbol{W}^* = Arg \min J(\boldsymbol{W})$.

An insufficient number of training input/output pairs relative to the size of the weight tensor would give rise to *overfitting*, requiring special techniques, generically known as *regularization*, in order to trim the network, i.e., randomly remove connections, without compromising predicting efficacy [2, 5].

Optimizing the network involves the laborious and costly computation of the minus gradient $-\partial J\left(\boldsymbol{W}\right)/\partial \boldsymbol{W}$, which locally indicates the direction of the steepest descent in the multidimensional surface $J=J(\boldsymbol{W})$. An iterative gradient descent computation generating a fine-tuned weight updating $\boldsymbol{W} \rightarrow \boldsymbol{W} - h\partial J(\boldsymbol{W})/\partial \boldsymbol{W}$ should eventually lead to convergence to a local minimum of $J=J(\boldsymbol{W})$ when adopting a suitable learning step h. This parameter should be tuned to effectively escape local minima while avoiding overshooting in trying to reach the global minimum. Most gradient descent algorithms use an adaptive learning step during training, in accord with the constraints indicated. In practice, the gradient problem is approached by what is called the *stochastic gradient descent* method, whereby not all datapoints (input/output pairs) in the training set are used in each minimization step, but the gradient is approximated by an average over randomly chosen batches of datapoints in a tradeoff between accuracy and computational efficiency. Obviously, batch size and learning rate are correlated, so the more accurate the gradient estimation, the larger the learning step may be (a token of computational confidence). To achieve significant speed, the stochastic gradient descent computation may be massively parallelized by splitting up batches into multiple GPUs.

5.2 NEURAL NETWORKS AS DYNAMICAL SYSTEMS

In this section, we shall be concerned with data organized as a time series arrayed as $\{\boldsymbol{x}(t_0), \boldsymbol{x}(t_0+\tau), \boldsymbol{x}(t_0+2\tau), \ldots, \boldsymbol{x}(t_0+L\tau)\}$, where $\boldsymbol{x}(t)$ is the vector of observables at time t, and τ is the interval that determines the time coarse-graining inherent to the sequential detection registered in the vector $\boldsymbol{x}$. The time series enables training of the NN such that the output vector $\boldsymbol{y} = \boldsymbol{y}(\boldsymbol{x}(t), \boldsymbol{W})$ should approximate $\boldsymbol{x}(t+\tau)$ when the input is $\boldsymbol{x}(t)$, and this correspondence is carried over all t in the training time series. Thus, for a given network architecture, the optimal network is the one that realizes the minimum of the loss function:

$$J\left(\boldsymbol{W}\right) = \sum_{n=1}^{L} \left\| \boldsymbol{x}\left(t_0 + n\tau\right) - \boldsymbol{y}\left(\boldsymbol{x}\left(t_0 + (n-1)\tau\right), \boldsymbol{f}, \boldsymbol{W}\right) \right\|^2 \tag{5.1}$$

Obviously, for a fixed activation function, a model for the time series may be given simply by $Arg \min J(\boldsymbol{W})$, but such a model would lack universality as it would be extremely parametrized, most likely over-parametrized, and would not prove insightful in the sense that it is not parsimonious. The discussion prompts us to inquire what truly constitutes a model. This question will be addressed subsequently.

The most common time series that humanity has collected since time immemorial stems from astronomical observation. For about two millennia until the time of Copernicus, and Newton later on, humanity had been striving to find a suitable model that would fit and explain the data, i.e., the recorded sequential positions of a set of celestial bodies. In today's more general context, biology, particle physics, and cosmology are generating dynamical data at a staggering rate, while models that fit and explain the data are sorely lacking or hopelessly inconsequential. It is expected that the advent of AI will dramatically impact this sort of model discovery.

In essence, we seek what is known as autoencoder, an intermediate output with a dimensionality reduction and a simplified discerning physical picture that should therefore prove insightful to make sense of the patterns enshrined in the dynamics, enabling meaningful output inferences. These autoencoders and the models they give rise to will be studied in detail in the subsequent chapter.

5.3 DEEP LEARNING AND CONVOLUTIONAL NEURAL NETWORKS FOR FEATURE EXTRACTION

The huge output of biomedical and biostructural data has become the hallmark of the post-genomic era, while chemical combinatorial possibilities make it forbiddingly difficult to parse chemical space in search of suitable leads for targeted therapy [6]. In this scenario, pharmaceutical researchers have turned to DL for guidance in drug discovery and development and target validation [7–10]. Thus, pharmacoinformatics has benefited immensely from the advent of DL systems trained to pair chemical compounds with molecular attributes likely to have a therapeutic impact. These computational and informational techniques enable the evaluation of chemical compounds for specific properties, including target affinity, affinity screening profiles, structural features of drug-target docking, and ADMET (absorption, distribution, metabolism, excretion, and toxicity) profiling [8].

A major challenge in implementing DL models for pharmacoinformatics in accord with the generic scheme outlined stems from the need to represent the chemical structure as a tensor array (vector, matrix, or tensor proper) of pixelated or voxelated inputs that may be subsequently interrogated geometrically across the hidden layers of NN in search for features that are indicative of the molecular properties indicated. There are a number of representations of chemical space amenable to TF encoding. The most obvious one is to order the atoms in the compound on a 1D-array (following, for example, the IUPAC numerical labeling convention) and represent the chemical structure of the molecule as a covalent bond matrix pairing atoms in row and column in accord with their covalent linkages, including single, multiple, and resonant (aromatic) bonds. The matrix is subsequently transformed into a topological descriptor that describes the invariants arising from different atom ordering. The input describing chemical structure is paired within a training set against the molecular attributes of therapeutic relevance that the network is meant to infer. Then, compounds that need to be evaluated/profiled are inputted as the array of pixels/voxels, transformed into a topological representation, and feature extraction is achieved through the sequential activation of hidden layers at increasing levels of abstraction, eventually leading to the profile inference which is subsumed in the output layer.

Feature extraction through the NN often requires particular architectures known as convolutional NNs (CNNs) [2–5]. The idea is to pixelize or voxelize the input data in a matrix or 3D tensor array and then scan (convolve) the array with a filter associated with a specific pattern to generate a feature map. For the sake of the argument, let us consider a 2D-array input $\boldsymbol{M}$. The filter $\boldsymbol{F}=(w_{ij})$ is an $m{\times}m$ matrix actually representing a convolutional kernel, so that a neuron in the F-associated hidden layer $\boldsymbol{M}_F$ only senses the pixels in an $m{\times}m$ patch (receptive field), and the layer becomes the feature map

$$\boldsymbol{M}_F = \boldsymbol{M} * \boldsymbol{F} = \left(\sum_{j=1}^{m} \sum_{i=1}^{m} w_{ij} x_{i+ap\ j+bp} + w_0 \right)_{ab}, \tag{5.2}$$

where p, usually set at p=2, is the stride adopted as the filter slides along the input matrix array, with dummy integers a and b indicating the patch location. Essentially, the CNN is an NN where the set of weights in each $m{\times}m$ patch of inputs is always the same as the filter slides along the input array to generate the hidden layer that constitutes the feature map. The convolution operation becomes the entry-by-entry (Frobenius) matrix inner product as the filter slides along the input matrix with a given stride. Successive filters may be applied reducing progressively the size of the feature maps as higher and higher levels of abstraction are achieved in the representation of the data (Figure 5.2). Thus, the parametrization required for feature extraction in CNNs is relatively small, as all the neurons in a hidden layer share the same connectivity parameters that define the filter that generated that layer.

In CNNs the filters are not specified *a priori*, i.e., their weights are not fixed through human intervention. The filter parametrization is automatically determined by the training/optimization process without introducing any assumption, other than the size of the receptive fields for each convolution operation and the overall number of filters to be applied. Thus, feature extraction in a CNN is carried out in an unsupervised manner and only requires that we script (in TF coding) the number of hidden layers or feature maps

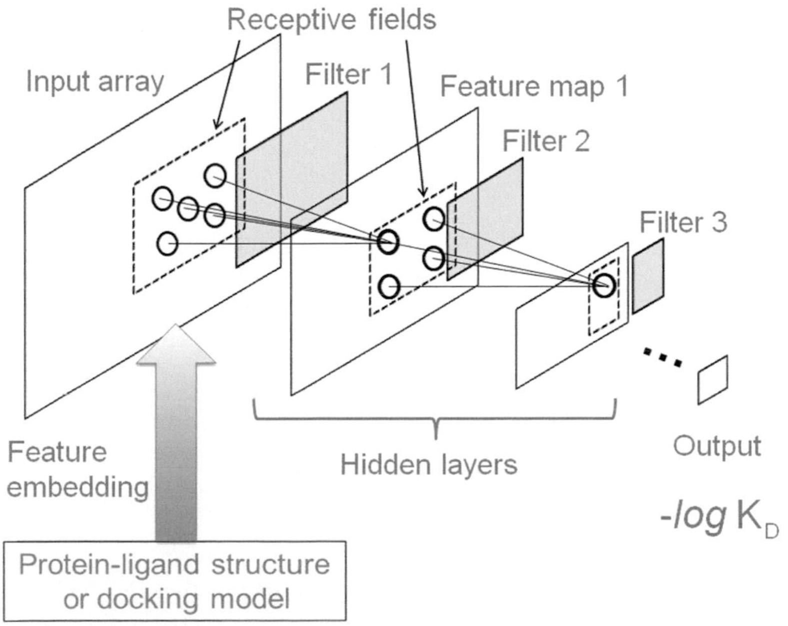

FIGURE 5.2 Scheme of a CNN used for inference of drug-target affinity through the sequential application of filters that become optimized through minimization of the network loss.

and the size of the filters to be applied to generate each hidden layer. The features themselves emerge as the network is trained.

Thus, CNNs often constitute AI-empowered platforms for drug discovery, where the structure of a protein-ligand interface that serves as a precursor for a predicted drug-target interaction is pixelated in a feature embedding process as a 3D spatial array of protein-drug atom pairs deemed to be interacting across the interface [11]. The inference of target affinity for a given drug is assessed through a sequence of feature extractions using convolution filters until the output feature map becomes a number directly associated with the affinity $pK_D = -\log K_D$, where K_D is the dissociation constant for the drug-target complex inputted via feature embedding (Figure 5.2). The training of the network is carried out by minimizing the network loss or empirical risk over a set of drug-target complexes whose structure and affinity are both known (preferably, experimentally determined). The proteins in the complexes of the training set are typically homologs of the one whose affinity for a specific drug we seek to infer, so that the features that enable the affinity inference may emerge from structural alignment.

In other more complex applications of CNNs, where the output is not simply a numerical parameter but rather a numerical array, it is often convenient to adopt a variation of the CNN architecture in which the patches that yield the pixels with the highest weights in the feature map are reconstituted into features perfectly stenciled by the filter, while other patches that do not yield discernible features are left invariant in non-overlapping regions. Thus, with the application of each filter, a feature-enriched reconstitution of the input layer is carried out prior to the application of the next filter (Figure 5.3).

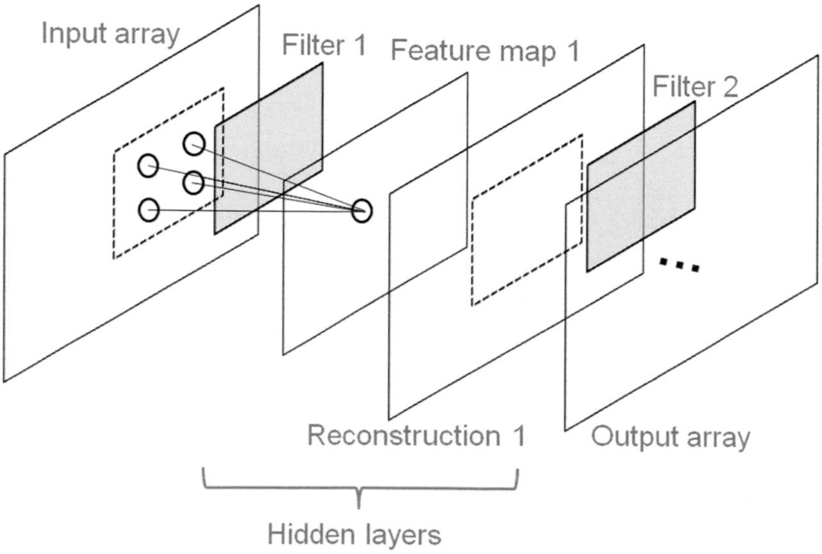

FIGURE 5.3 Scheme of a CNN with a particular type of architecture in which the patches that yield the pixels with the highest weights in the feature map are recreated as filter-stenciled "perfect" features. Thus, a feature-enriched reconstitution of each layer is carried out prior to the application of the next filter.

5.4 METAMODELS FOR ADIABATIC REGIMES, LATENT MANIFOLDS, AND QUOTIENT SPACES

It is widely expected that AI and DL in particular will become major players in model discovery for data-driven research [12]. Within this vast array of possibilities, the focus of this book is model discovery for dynamical systems that underlie big time series data from vast areas as distant as biomedicine and cosmology [13–15]. The biggest hurdle we stumble upon is that the level of complexity of the data generated in fields like biology and cosmology is so extreme that it challenges the notion of the model itself. As shown subsequently, the model itself can seldom be cast as a system of differential equations. Thus, much of the ensuing discussion deals with the question of what constitutes a meaningful model with predictive value when dealing with the ultra-complex realities represented in the contexts of biology or cosmology.

Encoding has proven to be a necessary category in AI [12–15]. The type of AI we are mostly concerned with involves deep learning, which requires an encoding of the raw information that needs to be acquired and processed further to make meaningful inferences. Just like with human intelligence, the encoding problem stems from the core question: What is essential, and what it superfluous? The encoding problem becomes solvable when the system under scrutiny is hierarchical and the hierarchical structure is fairly obvious or at least discoverable. In the cases treated in this book, the structure of the data is always hierarchical, implying that the learning process admits a reductive approach represented by the encoding. When the network architecture is such that this process is automatically generated, we name the NN *autoencoder*.

We shall deal mostly with time-dependent data representing physical or biophysical systems, where detailed fast motions may be systematically averaged out, so the relevant information may be stored in a coarse-grained representation. Rigorous mathematical constructs will be introduced to implement the hierarchical encoding materialized by the autoencoder. Some examples of hierarchical encoding of physical or biophysical processes that need to be taken into account when designing the autoencoder architecture are:

1. The adiabatic approximation, where fast-relaxing or fast-evolving enslaved degrees of freedom are averaged out, or thermalized or equilibrated when incorporated into a model for the time evolution of a dynamical system [16]. In molecular physics, examples of such degrees of freedom are vibrational hard modes that evolve on timescales of the order of picoseconds to nanoseconds, while soft modes evolve on longer timescales ranging from submicroseconds to seconds.

2. In atomic physics, the Born-Oppenheimer approximation represents regions of the potential energy surface where an adiabatic regime holds so the motion of electrons is enslaved or entrained by the slower motion of the atomic nuclei [17].

3. The latent manifold in dynamical systems [12] is where the system is entrained or enslaved in the long time limit by the evolution in a manifold of a lower dimension

than the original space. The latent manifold is often referred to as the center manifold [16], especially in the context of dissipative systems, and contains the attractors of the system (Figure 5.4).

4. In the biophysical context of functionalized soluble proteins (enzymes), the quantum mechanics for specific chemical processes take place at the *epistructure* (solvent organization around the protein structure), where water is chemically functionalized [18].

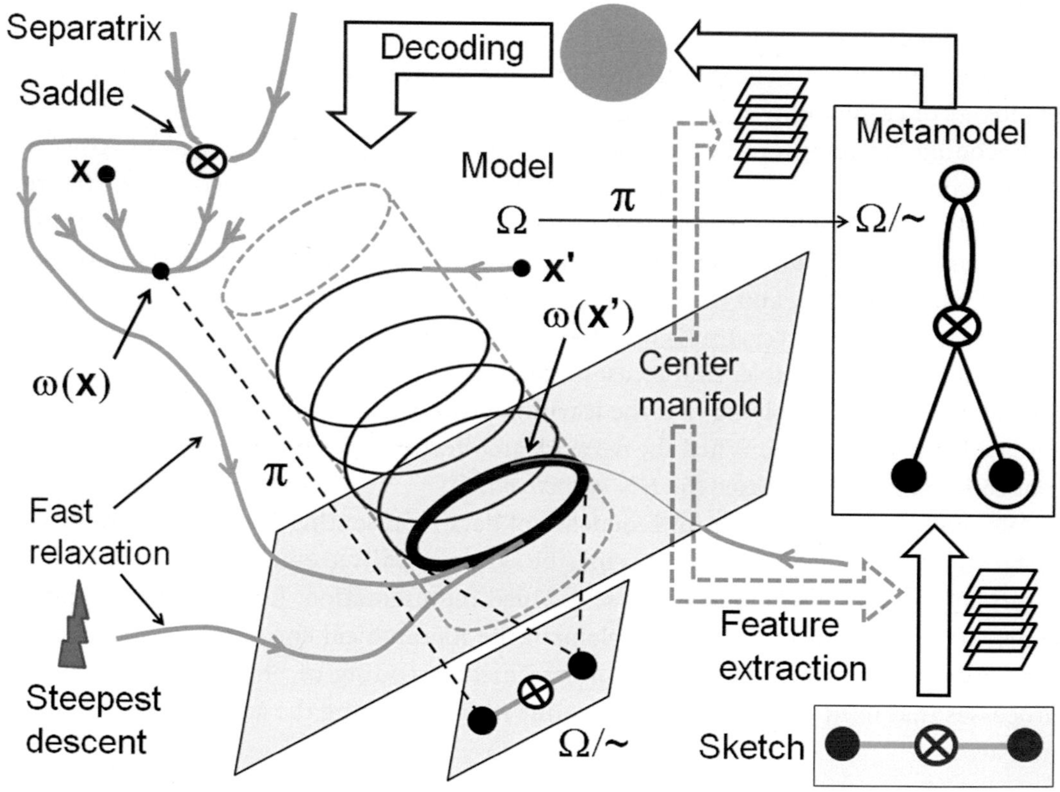

FIGURE 5.4 Schematics of the topological "metamodel" encoding of a dynamical system that contains a center manifold. The center manifold enslaves or entrains the dynamics for timescales associated with the adiabatic elimination of fast-relaxing and thermalized degrees of freedom and therefore constitutes a latent manifold (Ω) within which a model (differential equations on latent coordinates) may be identified by an autoencoder. The dynamic information may be encoded further by a second autoencoder at a higher level of abstraction, where the dynamics are represented more coarsely as "modulo-basin" transitions. The modulo-basin dynamics is mapped on a quotient space, Ω/~, where two states x and y are regarded as equivalent, x~y, if they have the same destiny state ($\omega(x)=\omega(y)$). The encoding processes are symbolized by dashed lines, and the enslaving center-manifold dynamics are highlighted by a thick dark circle. Thus, the second autoencoder materializes the projection $\pi: \Omega \rightarrow \Omega/\sim$ and represents the dynamics as a walk in a graph whose vertices are the critical points (minima, saddles of different indices, maxima) and attractors of the system, and the edges represent connections along pathways of steepest descent. This topological metamodel may be subsequently decoded back to flesh out the dynamical system using learning technology.

A rigorous treatment of this problem would require that we solve a time-dependent Schrödinger equation. In this context, an AI approach would need to be incorporated to learn to infer the nodal structure of molecular orbitals within a voxelated 3D grid. With current technologies, this AI approach is plausible only under the adiabatic regime given by the Born-Oppenheimer approximation.

5. In clonal population dynamics, cancer phenotypes become selected under therapeutic pressure [19]. These dynamics are complex but hierarchical, and the mathematical procedure to "encode what is essential" in this context is based on the center-manifold reduction (Figure 5.4). The reduction discussed in (3) is actually a projection of the dynamical system onto a lower dimensional system that entrains it. The autoencoder may discover the center manifold by interrogating vast amounts of time-dependent data, as it seeks to meaningfully reduce dimensionality in such hierarchical systems.

6. Chapter 6 introduces the *quotient space*, a fundamental mathematical construct to take advantage of dynamic hierarchy in order to encode information as required to implement DL systems [20]. The quotient space is built upon underlying dynamics and may be equated with the orbit space, i.e., points in the same trajectory are regarded as equivalent, and the quotient space is the set of equivalence classes with a topology inherited from the underlying space where the dynamical system is mapped (Figure 5.4). In rigorous terms, two points-states $(\boldsymbol{x}, \boldsymbol{x}')$ are equivalent $(\boldsymbol{x} \sim \boldsymbol{x}')$ or belong to the same equivalence class $(\boldsymbol{x}' \in \overline{\boldsymbol{x}}, \overline{\boldsymbol{x}'} = \overline{\boldsymbol{x}})$ if and only if they share the same destiny state $(\omega(\boldsymbol{x}) = \omega(\boldsymbol{x}'))$ vis-à-vis the trajectory to which they both belong. Thus, we simplify the space by lumping microstates into basins of attraction (of destiny states) in the potential energy surface. The modulo-basin dynamics constitute a coarse model named *metamodel*, which is far easier to encode as the system learns data generated by atomistic molecular dynamics simulations. In this way, AI learns to propagate dynamics in quotient space, discovering a metamodel to cover physically realistic timescales usually inaccessible to detailed atomistic computations.

While the center-manifold encoding is suited for dissipative dynamics, where the attractor may be nontrivial (cf. Figure 5.4), the quotient space simplification is better suited for Hamiltonian systems. Both levels of encoding converge as free-energy dissipation tends to zero. In fact, as the AI system encodes time-dependent raw data, it implicitly composes the two levels of encoding, with the center-manifold reduction averaging out of fast modes, followed by projection onto quotient space (Figure 5.4).

In all cases dealt with in this book, the hierarchy of the data that makes it amenable to encoding is either AI-discoverable (Appendix) or it may be unraveled through a rigorous mathematical construct that needs to be incorporated into the learning code, as shown below.

The modulo-basin hierarchical representation of the dynamics in quotient space enables the construction of metamodels, that is, coarse-grained models that represent transitions

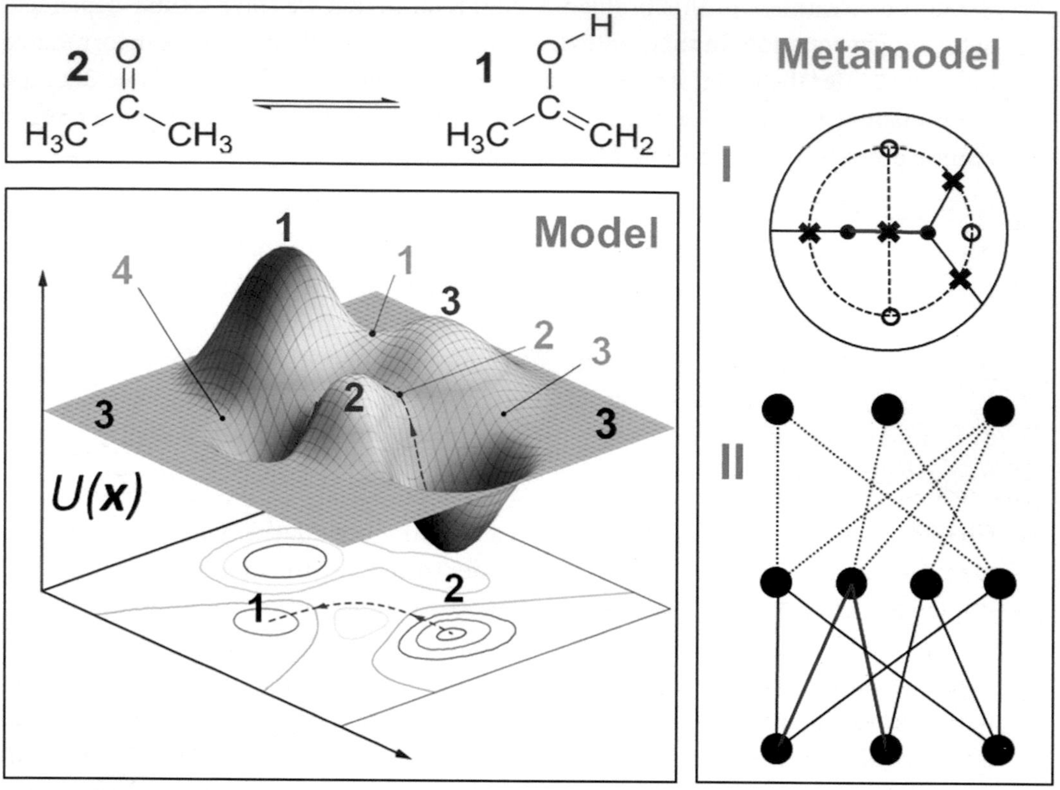

FIGURE 5.5 Three levels of abstraction in the modeling of the chemical dynamics for the isomerization "keto" ⟺ "enol" of acetone, representing the intramolecular migration of a proton. The potential energy surface (PES) represents the model and the topological relations between critical points, together with the "modulo-basin" transitions along the paths of steepest descent joining critical points represent the metamodel. On metamodel I, the modulo-basin topology is mapped onto the latent manifold, while on metamodel II, the modulo-basin topology is abstracted further and displayed as a graph.

between equivalence classes of states of the system, essentially reproducing the topological dynamics within a graph representation (Figure 5.5). As an illustration, let us consider the "keto" ↔ "enol" interconversion of acetone (Figure 5.5). This is a chemical reaction involving an intramolecular proton migration that can be modeled by the potential energy surface (PES) representing the ground-level electronic energy sheet under the Born-Oppenheimer approximation. From a quantum mechanics calculation, we know that this PES has three minima, representing the less stable "enol" form (1), the more stable "keto" form (2), and a state where the jumping proton is completely dissociated from the rest of the molecule (3). The minima are separated by four saddle points at the top of the path of steepest descent joining pairs of minima on both sides of the saddle along the direction of negative curvature. This direction is actually the eigendirection associated with the eigenvalue of the Hessian matrix (Jacobian of the gradient vector field) with a negative real part computed at the saddle point. In turn, the lines of steepest descent joining the four saddles with the three maxima become the separatrices of the basins of attraction of the minima.

Thus all lines of steepest descent may be represented as a graph (metamodel II, Figure 5.5) joining critical points that are in turn organized in tiers, where the lowest tier corresponds to points where all eigendirections have positive curvature (minima) and the next tier includes all critical points with only one eigendirection with negative curvature (saddles of index 1), the next tier is associated with two eigendirections of negative curvature (maxima in the case of Figure 5.5), etc. The edges on the graph represent paths of steepest descent joining critical points in adjacent tiers. In this way, the topological dynamics of the PES may be encoded as a metamodel represented in graph form, while the reversible chemical reaction pathway becomes a walk in the graph.

As argued subsequently, metamodels enable the discovery of hierarchical dynamical systems underlying processes that unravel in realities of high multi-scale complexity. The implementation of a metamodel factorization of the dynamics requires the concerted participation of several components operating in a coordinated manner within an AI platform. First, we need to introduce the so-called autoencoders that constitute the deep NN systems that encode the dynamics on the "latent manifold" Ω. This manifold entails a significant dimensionality reduction relative to state space W and is spanned by the internal coordinates that label the orbits of the symmetry group $\mathcal{G}$ inherent to the system and acting on W: $\Omega \approx W/\mathcal{G}$. Traditionally, it is expected that the first autoencoder, hereby denoted AE1, which generates the latent manifold, is jointly optimized to generate also the most sparse or minimal set of differential equations on the manifold that can be decoded back onto the dynamical system defined on W. This is what is usually meant by "model discovery." In practice, the level of multi-scale complexity of the processes dealt with in this book does not make models amenable to discovery at the geometric level. As previously argued, in such cases, another level of abstraction π: Ω→Ω/~ needs to be introduced so that the coarse-graining of time within Ω/~ reflects equilibration within the basins of attraction in the latent dynamics. This hierarchical escalation in the level of abstraction requires a second autoencoder, AE2, capable of encoding the topological features of the latent vector field.

At this stage, a different sort of NN architecture is required to propagate the metadynamics on Ω/~. To properly delineate the architecture of the NN required for metadynamic propagation, we limit the discussion to the case where the dynamics are generated by a smooth (i.e., C^1) potential energy function $U: W \rightarrow \mathbb{R}$ invariant upon the isometries (distance-preserving transformations) of W. Furthermore, $\mathcal{G}$ is a Euclidean group, so that Ω is compact, and hence Ω/~ becomes a discrete set of basins, and a basin assignment represents a coarse state of the system. Then, the metadynamics may be encoded as "evolving text" representing a time series of basin transitions. The application described in the Appendix is primarily devoted to constructing and exploiting this particular type of metamodel. The textual processing requires the implementation of a particular type of DL architecture known as a *transformer*, while the transformer-based propagation of the metadynamics constitutes a Markovian process (Figure 5.6). Through AE2, this metadynamics is decoded as latent dynamics on Ω and validated by contrasting the latent dynamics against the hidden Markov process upheld under the adiabatic conditions described.

Thus, to implement a metamodel within an AI platform, it is essential that the two autoencoders and the transformer are optimized to work concertedly with complete

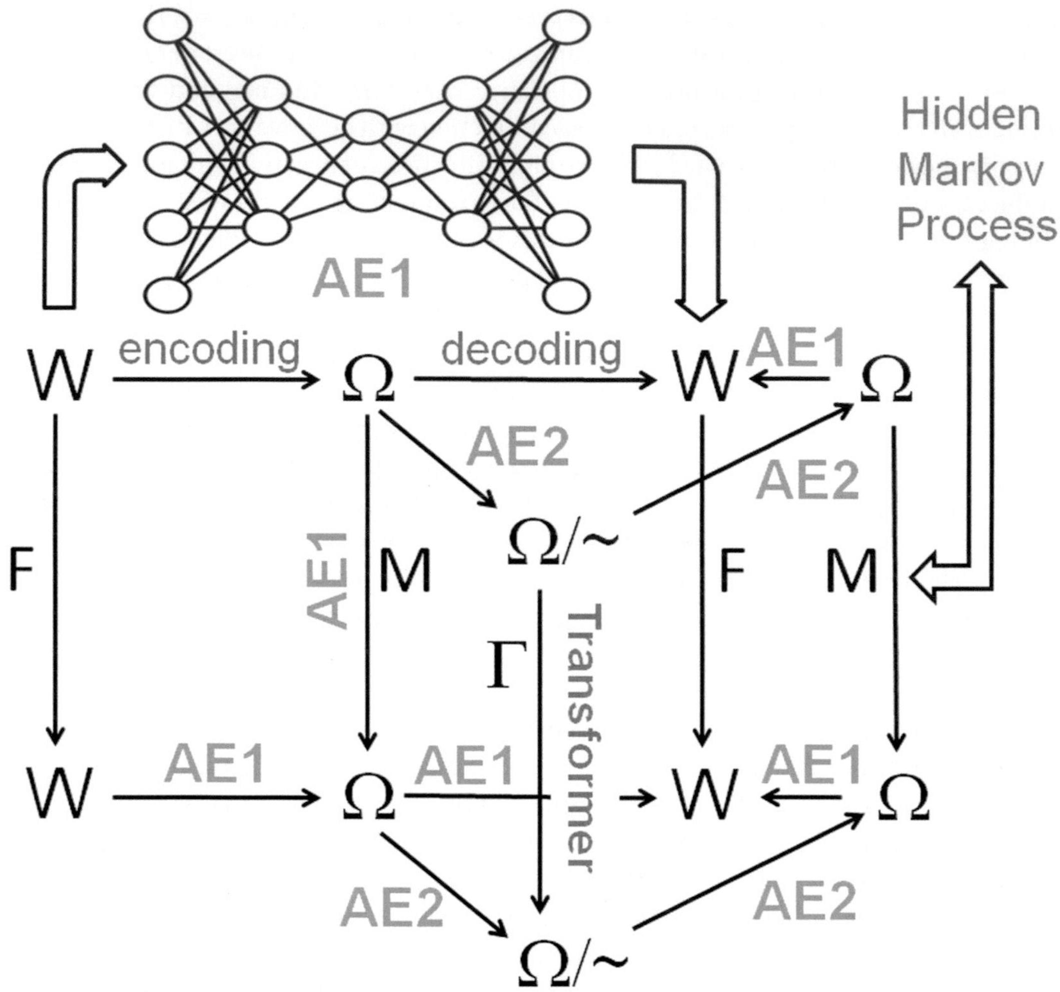

FIGURE 5.6 Commutative diagram representing two coupled autoencoders operating sequentially in tandem and representing two levels of abstraction of a dynamical system. The first autoencoder, labeled AE1, projects the dynamics onto the latent manifold, while autoencoder AE2 projects the latent dynamics onto a discretized "modulo-basin" version, where a coarse state of the system is represented in textual form, and the "modulo-basin" dynamics may be learned and propagated within a special type of auto-encoding architecture known as "transformer."

compatibility (Figure 5.6). This means that an input state yields the same destiny state regardless of the pathway chosen provided the pathways have identical endpoint spaces in the commutative diagram presented in Figure 5.6.

5.5 THE STANDARD MODEL AS A DYNAMICAL SYSTEM

The notion of diagram commutativity is central to the encoding of a dynamical system into its latent dynamics and, reciprocally, to the decoding of the latent dynamics into the full dynamics. Thus, the task of decoding the Standard Model, canonically defined on the four-dimensional space-time W/~, onto a five-dimensional compact multiply connected manifold W requires an autoencoder operating in reverse. This reverse autoencoder lifts the flow $\mathcal{F}: W/\sim \rightarrow W/\sim$ to a flow $\tilde{\mathcal{F}}: W \rightarrow W$ that must be compatible in accordance with the commutativity condition: $\mathcal{F} \circ \pi = \pi \circ \tilde{\mathcal{F}}$, with π: W→W/~ denoting the canonical projection that assigns each point in W to its equivalence class in W/~. The key issue with this formulation is that the Standard Model does not represent a dynamical system in any obvious way, so what sort of flow are we actually discussing in this context? This question is crucial because we are introducing AI technology that is tailored for dynamical systems.

Fortunately, we may treat the interaction processes described by the Standard Model as transformations created by a time-dependent propagator. For each elementary process, this operator has a time step associated with it, and this time step is exactly the lifetime of the boson that communicates the force in the process of particle transformation, as obtained from Heisenberg's uncertainty principle: $\sim \hbar/2M_Bc^2$, where M_B is the rest mass of the respective gauge boson.

Thus, the AI technology described in this chapter will be applied in reverse in order to reverse engineer the Standard Model by incorporating a dormant dimension shown to store gravity.

5.6 AI-BASED METAMODEL DISCOVERY FOR DYNAMICAL SYSTEMS

With the leveraging of artificial intelligence (AI) [1] and in particular, machine learning approaches [2, 3], dynamical systems have found a new fertile ground for further development [12, 13]. Showcase problems in applied mathematics, including the Lorenz strange attractor, reaction-diffusion spatiotemporal systems, and fluid dynamic flows captured by the Navier-Stokes equation, are being examined in a new guise as autoencoders identify parsimonious models with reduced dimensionality [14]. Such models are meant to provide the physical underpinnings of the phenomena enshrined in time series data or generated by raw differential equations.

Machine learning or more broadly, AI, is being leveraged for model discovery of dynamical systems underlying data represented as a time series. The data-regression system, which in this context is a neural network predicting future behavior, is trained by the time series and regarded as the model itself. However, this model is heavily parametrized and hence too "fragile" to allow for extrapolation [12]. In other words, such models are not really amenable to yield physical laws, the way other data-regression approaches are [15, 21]. This statement has been voiced repeatedly and hints at some level of dissatisfaction: Machine learning, and AI in general, are very efficient at providing predictive models when trained on a sufficiently long time series but often do a poor job at providing physical insights regarding the underlying dynamical system.

This problem gets significantly amplified as we turn to biological or, more broadly, biomedical matter [19]. It is widely felt that, when examined in their multi-scale richness and complex heterogeneity, dynamical systems in biology or biomedicine cannot reach the level of maturation required to be subsumed into applied mathematics. This statement should be interpreted in the sense that we lack sparse enough models that provide physical underpinnings of biological/biomedical phenomena and are suitable for extrapolation.

Will the leveraging of alternative AI-based approaches change the *status quo*? This chapter portends to address this problem and provide insights that will be methodologically fleshed out in the subsequent chapters to enable metamodel discovery by reverse engineering time series stemming from highly complex multi-scale realities.

A key question in fostering the mathematical maturation of model discovery in biology may be cast as follows: What constitutes a parsimonious model that provides physical underpinnings of biological phenomena? A standard answer with broad bearing on most problems considered tractable in applied mathematics is: "a sparse system of differential equations on a smooth manifold of latent coordinates" [14, 21]. As this chapter argues, this may be simply too much to ask for in the context of biological matter. Furthermore, this is not necessarily the format or framework that AI would typically choose for model discovery, given the "dimensionality curse" associated with the molecular reality of biological systems [19]. In principle, a single autoencoder that optimizes for sparsity in the discovery of the latent manifold might not provide a satisfactory solution to the modeling problem. The molecular reality *in vivo* typically has well over a million coordinates required to specify the state of the system, and the extent of connectivity parametrization for an autoencoder capable of handling such levels of complexity would be simply enormous, implying that the training and variational optimization of the neural network would be off-limits, at least with current computational capabilities [22].

This chapter squarely addresses this matter. To do so, it leverages AI methods to circumvent the difficulties associated with model discovery for time-dependent phenomena arising in soft or biological matter. In essence, as we shall show, AI dwells on a paradigm of what constitutes a parsimonious metamodel that is significantly different from the one adopted by applied mathematicians [19]. Thus, to identify the most economic yet faithful metamodel, AI will be shown to use two or more tandem autoencoders instead of one, as is typically done in model discovery [14]. The autoencoders are coupled and become fully compatible with each other at the completion of the parameter optimization process, as defined precisely in the subsequent sections. In the simpler cases where two autoencoders are required, the second autoencoder translates the dynamic information embossed in the latent manifold, turning it into a topological dynamics metamodel [23], which can be decoded and enables significant propagation of the dynamics into the future. This property is essential for coverage of realistic timescales relevant to the level of state extrapolation required. Crucially, the topological dynamics metamodel constructed by AI is essentially a pattern-based model, not a system of differential equations, as would be expected for Standard Model discovery in dynamical systems. This does not mean that AI is discovering laws without equations but simply that AI adopts a different way of casting models susceptible to extrapolation as recognizable physics laws. Thus, AI may not straightforwardly give us the equations that govern *in vivo* protein folding, but the underlying physics discovered may in all likelihood be cast in terms of AI-interpretable topological patterns that signal the commitment of the chain to fold into a steady conformation [20].

These topological dynamics methods are likely to advance the field of AI-based model discovery, as they enable the reverse engineering of time series data stemming from vastly complex hierarchical realities. Such contexts are illustrated for example by the cellular setting that assists and expedites molecular processes, which have been hitherto considered off-limits to machine learning approaches to dynamical systems.

At this stage of development, the topological methods introduced yield AI-recognizable patterns but do not beget latent differential equations that have traditionally been the hallmarks of model discovery. This may be viewed as a limitation in some sense, but we argue that that assertion is perhaps a reflection of narrow-mindedness. Synergistic efforts involving AI are likely to dominate future human endeavors in science, and AI systems are very much attuned to encoding and processing patterns in metamodels such as those produced by the topological approaches introduced in this book.

The impact of the topological methodology is likely to be broad, as it would render tractable problems in dynamical model discovery that have been hitherto considered off-limits by applied mathematicians that are currently incorporating machine learning in their toolbox. Thus ultra-complex realities recreating cellular environments that influence and steer molecular dynamics, such as those treated in Chapters 3 and 4, are likely to be within reach as topological methods are incorporated into the AI-empowered metamodel discovery.

5.7 AUTOENCODERS IN THE QUEST FOR LATENT COORDINATES

Since the dawn of modern science, Western civilization upheld the belief that understanding the workings of the universe pivots to finding fundamental equations that govern physical processes. In the case of a dynamical system, the underlying differential equations represent the basic model assumed to enshrine the physical laws that underpin the process described. The specific constraints, conservation principles, and symmetries of the system must all be taken into account when positing the differential equations. After centuries of work within this paradigm, it would be interesting to see how the leveraging of AI in synergy with the human endeavor will affect the choice of the format within which the physical laws are encoded and how this choice impacts the field of dynamic modeling. With the exponential development of computer technologies [22], we may soon be witnessing a paradigm revision. Equation-based modeling may or may not remain the dominant paradigm. Other types of dynamic descriptors are already making strides and contributing to a new understanding of the universe, or, at least, of the dynamic multi-scale complexities of reality [19]. These descriptors are certainly different and arguably more pliable than what biological humans managed to achieve so far.

The discovery of physical laws distilled from sequential data representing a time series remains a major challenge as well as an imperative to enhance our understanding and control of physical processes. In recent times, this type of data-driven model discovery has been fueled by significant breakthroughs [15, 19, 21, 24]. Yet, we also live in an era of big data, especially stemming from fields like biology and astrophysics, where the multi-scale complexity of the data organization makes model discovery particularly daunting. It is unclear whether the enormous richness of the data describing *in vivo* contexts at the molecular level is amenable to the kind of parsimonious models based on differential equations that mankind holds dear and has sought since time immemorial. More than in any other field, in biology it is likely that a willy-nilly application of Occam's razor may lead to self-inflicted wounds.

Deep learning (DL) approaches realized through autoencoder architectures have proven particularly valuable for the discovery of data-driven models represented by differential equations framed on "essential coordinates." The latter, often referred to as "latent coordinates" [21], span the so-called center manifold in dynamical systems [16, 19]. This reduction entails a significant dimensionality reduction and usually identifies the enslaving "slow" process that dynamically subordinates fast-relaxing modes [12, 16] and serves to encode the physical process we seek to model. Latent coordinates need to be selected very carefully, not only by weighing the extent of dimensionality reduction and the compactness and smoothness of the latent manifold but also by the economy of the resulting model. This economy is typically assessed by the complexity of the differential equations in latent space, quantified by the number of nonlinear terms [14].

In generic terms, an autoencoder seeks to identify latent coordinates **x** as an output of a feed-forward neural network (NN) with multiple hidden layers. The NN is inputted observable vectors, denoted generically as **z**, that serve to train the network, as the decoding of **z** from the latent coordinates must reproduce the inputted **z**-value (Figure 5.7). The autoencoder is optimized variationally, meaning that the activation weight parametrization of the multi-layered encoding and decoding functions, denoted respectively γ and μ, minimizes the loss function $\mathfrak{L}(\gamma,\mu)$ that measures the efficacy in the recovery of the input vectors:

$$\mathfrak{L}(\gamma,\mu)=Q^{-1}\sum_{z\in\mathfrak{F}} \| z-(\mu\circ\gamma)z \|^2 ;\gamma,\mu=\arg min_{\gamma,\mu}\mathfrak{L} \tag{5.3}$$

where $\mathfrak{F}$ is the training set and $Q=|\mathfrak{F}|$ is its cardinal, for now, regarded as a hyperparameter fixed to avoid overfitting [2, 3, 12, 13] in accord with the dimensions of the NN.

The autoencoder is functionally operative to model the dynamical system that underlies the given time series $\{z(t_0+n\Delta t), z(t_0+(n+1)\Delta t)\}_{n=0,1,2,...,L}$ if and only if the following relations hold for $t=t_0+n\Delta t$ with $n=0,1,2,\ldots,L$ ($L>>1$) and any initial time t_0:

$$(K_\Delta\circ\gamma)z(t)=\gamma z(t+\Delta t)=(\gamma\circ F_\Delta)z(t);(F_\Delta\circ\mu\circ\gamma)z(t)=(\mu\circ K_\Delta\circ\gamma)z(t)=z(t+\Delta t) \tag{5.4}$$

where F_Δ, K_Δ are the infinitesimal time maps in state space W and latent manifold Ω, respectively (cf. Figure 5.7). In other words, as the autoencoder becomes variationally optimized, the functional relations $K_\Delta\circ\gamma=\gamma\circ F_\Delta;F_\Delta\circ\mu=\mu\circ K_\Delta$ hold.

Thus, diagram commutativity (Figure 5.7) becomes the key property that enables us to assert that the choice of latent coordinates $x\epsilon\Omega$ was the "right" one in the sense that it captures the entrainment of the dynamics on W by the reduced dynamics on the latent manifold Ω. The commutativity rules ensuring the compatibility of raw and latent dynamics may be cast in terms of derivatives using the chain rule [14] as Δt is taken to the infinitesimal limit dt.

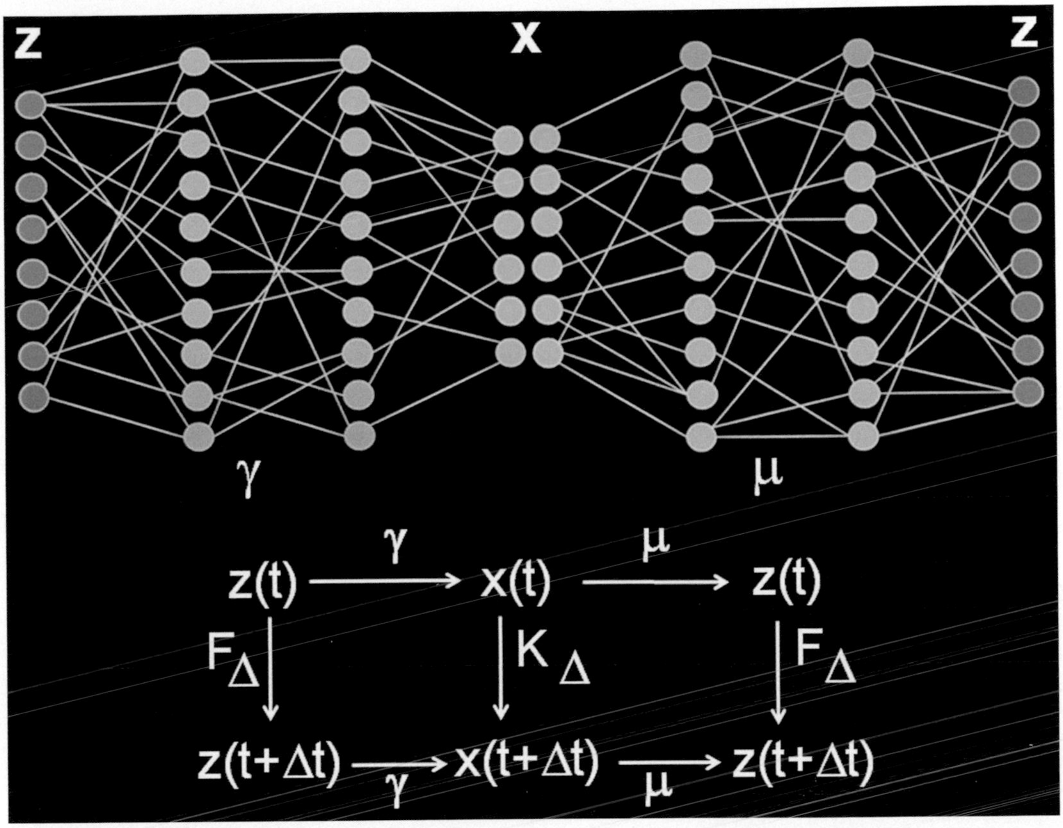

FIGURE 5.7 Generic scheme of the neural network (NN) architecture for autoencoder (γ,μ). The vector $z \in W$ represents the state of the system, and $x \in \Omega$ represents an encoded latent state that is decoded back into z. The encoding process entails a dimensionality reduction with $\dim\Omega < \dim W$ and, concomitantly, a coarse-graining of time in multiples of a time step Δ associated with the propagation of the dynamics on the latent manifold Ω. Gray disks represent nodes in hidden layers for encoder $\gamma = \gamma(\Theta)$ and decoder $\mu = \mu(\Theta)$, where Θ denotes the weights of node connections that are variationally optimized according to a loss function $\mathfrak{L} = \mathfrak{L}(\Theta)$. The weights realizing the variational minimum ($\Theta = \arg\min \mathfrak{L}$) in the training of the NN are optimal at making the diagram commutative.

5.8 MODEL DISCOVERY WITH DEEP LEARNING

Model discovery in dynamical systems has been maturing for some time, to the extent that it has been integrated into the corpus of applied mathematics. Furthermore, many applied mathematicians have incorporated DL into their toolbox as they make forays into reverse engineering of dynamical systems [12]. The overarching goal is to develop regression methods that enable the extraction of parsimonious dynamics from large data organized as time series $\{\mathbf{z}(t)\}_t$ with t given as multiples of a fixed time step [12, 14]. With the advent of DL, autoencoder architectures have successfully identified latent (essential) coordinate frames with significant dimensionality reduction ($\dim\Omega < \dim W$). Ultimately, the parsimonious description should become solvable in the form of a differential system $\dot{x} = K(x)$ that spans a differential system $\dot{z} = F(z)$ for the time series vector **z**. Yet a mere stark reduction in dimensionality does not necessarily ensure a parsimonious description [14]. The variational optimization of the autoencoder should simultaneously consider both the dimensionality of the coordinate frame for the latent manifold and the sparsity of the equations in the latent manifold, with the latter measured by the number of terms in the flow (vector field) that determines $\dot{z}$.

The autoencoder architecture and the required dual diagram commutativity between the latent flow **K** and the given time series flow **F** are jointly represented in Figure 5.8. We adopt a time step τ as the time series hyperparameter to discretize the flows by coarse-graining time resolution. The loss function $\mathfrak{L}(\gamma,\mu,K)$ now includes five terms: $\mathfrak{L}_r(\gamma,\mu)$, previously introduced, weights the efficacy of the recovery of the **z**-value upon encoding and decoding; $\mathfrak{L}_s(K)$ accounts for the sparsity of the latent flow **K**; and the terms $\mathfrak{L}_C(\gamma,K), \mathfrak{L}_C(\gamma,\mu), \mathfrak{L}_C(\gamma,\mu,K)$ represent the penalties associated with imperfect diagram

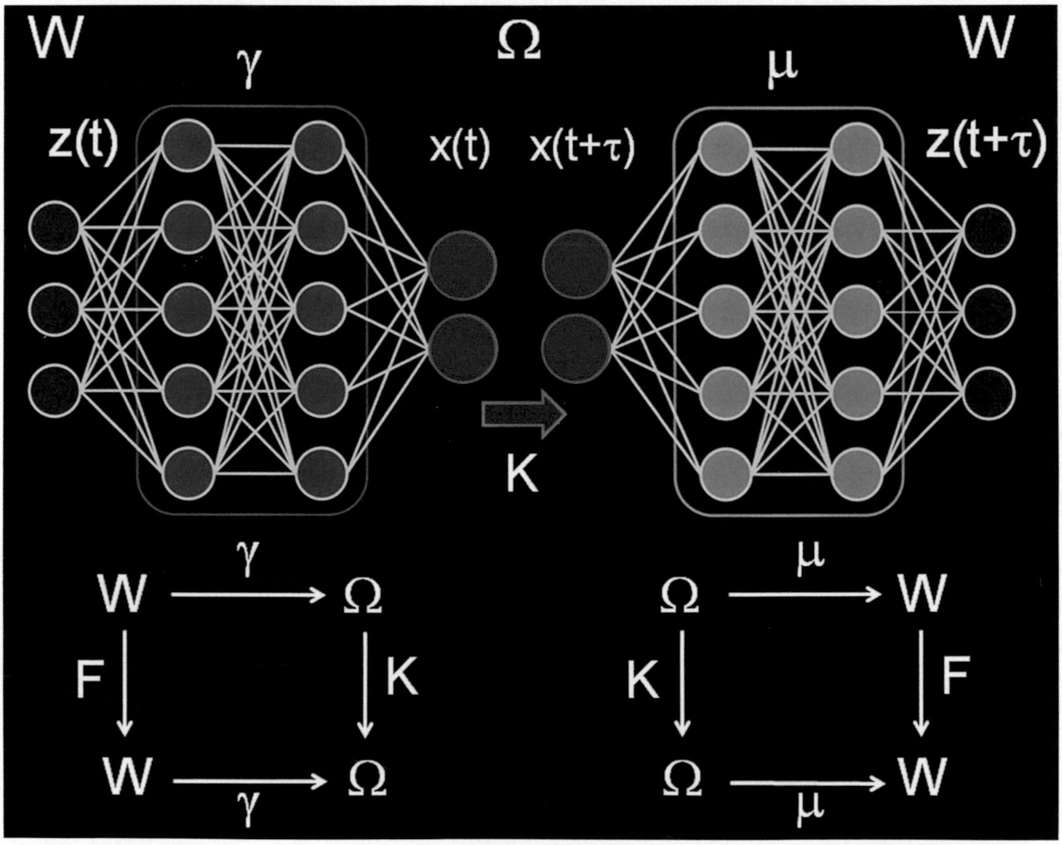

FIGURE 5.8 Scheme of autoencoder for latent dynamics optimizing simultaneously the dimensionality reduction and the parsimony of latent dynamical system governed by the flow K. The optimization is extended further relative to the architecture displayed in Figure 5.7 by adding the extra term $\mathfrak{L}_s(K)$ to yield the loss functional $\mathfrak{L} = \mathfrak{L}(\Theta, K) = \mathfrak{L}(\Theta) + \mathfrak{L}_s(K)$. The optimal Θ,K pair realizing the variational minimum is precisely the one that makes the bottom diagrams commutative.

commutativity that need to be imposed to guarantee the dynamic compatibility of the **z** and **x**-descriptions of the system. Thus, we get

$$\mathfrak{L}(\gamma,\mu,\boldsymbol{K}) = \mathfrak{L}_r(\gamma,\mu) + \vartheta_s\mathfrak{L}_s(\boldsymbol{K}) + \vartheta_C\left[\mathfrak{L}_C(\gamma,\boldsymbol{K}) + \mathfrak{L}_C(\gamma,\mu) + \mathfrak{L}_C(\gamma,\mu,\boldsymbol{K})\right], \tag{5.5}$$

where the relative weights ϑ_s, ϑ_C are hyperparameters and

$$\mathfrak{L}_r(\gamma,\mu) = Q^{-1}\sum_{z\in\mathfrak{F}} \| z - (\mu\circ\gamma)z \|^2, \tag{5.6}$$

$$\mathfrak{L}_C(\gamma,\boldsymbol{K}) = Q^{-1}\sum_{z\in\mathfrak{F}} \| (K\circ\gamma)z - (\gamma\circ F)z \|^2, \tag{5.7}$$

$$\mathfrak{L}_C(\gamma,\mu) = Q^{-1}\sum_{z\in\mathfrak{F}} \| (F\circ\mu\circ\gamma)z - (\mu\circ\gamma\circ F)z \|^2, \tag{5.8}$$

$$\mathfrak{L}_C(\gamma,\mu,\boldsymbol{K}) = Q^{-1}\sum_{z\in\mathfrak{F}} \| (\mu\circ K\circ\gamma)z - (F\circ\mu\circ\gamma)z \|^2, \tag{5.9}$$

with $Q = |\mathfrak{F}|$ indicating the cardinal of the training set.

To determine the sparsity term $\mathfrak{L}_s(\boldsymbol{K})$, we first take the limit case where τ becomes the time infinitesimal dt and assume that $\dot{\boldsymbol{x}} = \boldsymbol{K}(\boldsymbol{x}) = \boldsymbol{AP}(\boldsymbol{x})$, where $\mathbf{A}=[A_{ij}]$ is an $m\times p$ matrix ($m=\dim\Omega$) whose coefficients are variationally optimized jointly with the parametrization of the autoencoder maps γ and μ, and $\mathbf{P}(\mathbf{x})$ is a p-vector given by $\boldsymbol{P}(\mathbf{x})^T = (1\ x_1\ x_2\ x_3\ x_1^2\ x_2^2\ x_3^2\ x_1x_2\ x_2x_3\ x_1x_3\ x_1x_2x_3\ldots)$ consisting of p functional terms. The terms do not need to be of a polynomial form, and they may be chosen in accord with the type of symbolic regression. Thus, the variational optimization of $\boldsymbol{A}$ is tantamount to a symbolic regression in the latent manifold Ω. To ensure the sparsity of the model, we define the loss term as $\mathfrak{L}_s(\boldsymbol{K}) = \| \boldsymbol{A} \|_F$, where $\| . \|_F$ is the Frobenius norm given by

$$\| \boldsymbol{A} \|_F = \left[\sum_{i=1}^{m}\sum_{j=1}^{p} A_{ij}^2\right]^{1/2} \tag{5.10}$$

The incorporation of this term into the loss function given by Equation (5.5) ensures that the latent manifold is selected to provide a parsimonious model with reduced dimensionality governed by the simplest possible set of differential equations in consonance with the time series provided. A different but equivalent formulation of the autoencoder has been given elsewhere [14] and uses the chain rule of derivation rather than diagram commutativity to infer the proper loss functional.

5.9 EXTENDING MOLECULAR DYNAMICS WITH VARIATIONAL AUTOENCODERS

As applied mathematicians make their forays into combining dynamical systems with machine learning, they adopt specific systems that have traditionally represented showcases for model building. Examples of such systems are the Lorenz attractor generated by three coupled ordinary differential equations, the spatiotemporal reaction-diffusion systems, and the hydrodynamics and turbulence models governed by the Navier-Stokes equation [14, 24]. Seldom, if ever, have we had a chance to evaluate how autoencoder techniques pan out in the realm of molecular dynamics for ultra-complex multi-scale many-body systems. We are specifically referring to the discovery of models that distill the cooperative collective motion of ensembles of atoms, atomic groups, molecules, or molecular assemblages in condensed phases characterized as clusters, liquids, glasses, crystals, polymers, etc., under overarching categories such as soft matter and biological matter. The foundational approach to model discovery for such systems has been traditionally provided by statistical mechanics. These methods have met considerable success, except in the realm of biological matter, where problems such as the discovery of protein folding pathways or the role of in vivo/cellular contexts in expediting the folding process cannot be even remotely addressed due to their sheer complexity and heterogeneity. Interacting molecular units are simply too diverse, and the cellular environment is too complex at multiple scales and heterogeneous for statistical mechanics to make successful forays in biology [19, 25, 26]. This is precisely the context where AI may empower dynamical systems by leveraging highly specialized autoencoders and even batteries of autoencoders, as described subsequently.

To furnish a general framework, we may start by noting that we are dealing with N particles that may be free or tethered through covalent linkages forming assemblages such as biopolymers that may interact with one other through ephemeral or permanent associations that do not involve covalent bonds. A protein chain embedded in an aqueous environment and interacting with other biomolecular entities such as protein enzymes, chaperones, or ribosomes represents the quintessential situation that we wish to address in this book as we learn to leverage AI methods and integrate them into a model discovery platform.

Keeping, for now, the discussion at its most generic level, let us consider an ensemble of N particles that has associated with it an internal energy $\mathcal{U} = \mathcal{U}(\hat{z}), \hat{z} \in \mathbb{R}^{3N}$ only dependent on inter-particle distances and hence invariant upon isometries – distance-preserving maps – of $\mathbb{R}^3$. Then, the state or configuration of the system may be represented as a point z in the quotient space $W = \mathbb{R}^{3N}/E^+(3)$, where $E^+(3)$ is the special Euclidean group of isometries of $\mathbb{R}^3$ including only rigid-body translations and rotations but excluding reflections [27]. The latter are excluded because they do not preserve the chirality (handedness) of asymmetric tetrahedral carbon groups, which constitutes a key constraint: Chirality is known to be strictly preserved in biology. Thus, W is (3N−6)-dimensional, and its coordinates represent all the internal degrees of freedom of the system. In addition to the reduction modulo isometries, another quotient is required to represent the latent manifold Ω. This further dimensionality reduction depends on the constraints brought about by the covalent linkages that tether specific units in the system. For example, if we are studying the dynamics of protein folding, we note that high-frequency vibrational motions involving covalently paired atoms may be averaged out, as their associated timescales, in the femtosecond (fs) range [28], are incommensurably shorter than those associated with soft modes represented by dihedral torsions of the polymer backbone (Figure 5.9), typically in

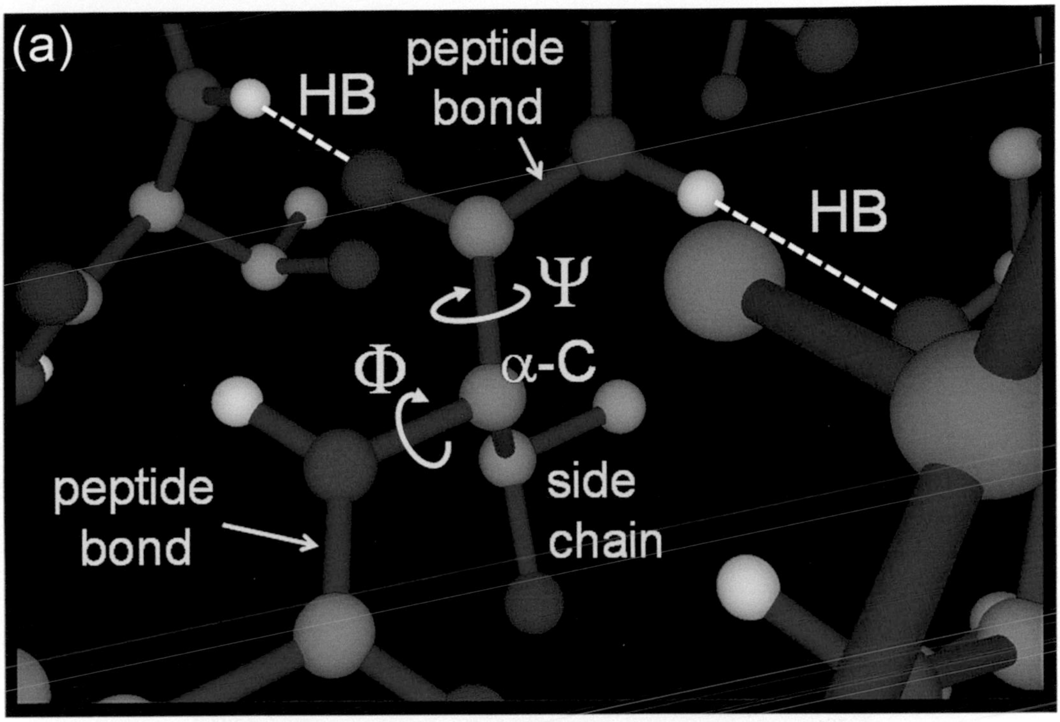

FIGURE 5.9 Latent coordinates for a folding protein chain. (a) Individual residue unit within a protein chain attached to an adjacent residue through a torsionally rigid linkage known as a peptide bond. For the sake of illustration, a particular protein has been selected, namely the thermophilic variant of the B1 domain of protein G from *Streptococcus*, whose native 3D structure is found at the entry 1GB4 in the protein data bank (PDB). In the latent coordinate frame, the state of each residue is represented by a pair of coordinates (Φ, Ψ) representing the torsional dihedral degrees of freedom of the protein backbone. The amino acid type for the residue, describing the local chemical composition of the chain, is identified by the side-chain group that is covalently linked to the alpha-carbon in the protein backbone. In turn, polar groups in the backbone and side chain can be engaged in orientationally and distance-dependent noncovalent linkages known as hydrogen bonds (HB, dashed lines), whose cumulative folding-stabilizing effect becomes a determinant of the protein 3D structure.

the nanosecond (ns) or sub-ns range. A similar reduction applies to planar angular vibrations, with frequencies in the order of $(\text{fs})^{-1}$ to $(10\text{fs})^{-1}$ [19, 28]. These simplifications point to a parsimonious model representing an adiabatic system that incorporates only soft modes of the chain. Thus, as protein folding is represented as a dynamical system with the protein chain searching in conformation space in an *in vitro* setting, Ω becomes a compact manifold in the form of a $2M$-torus, where M is the number of amino acid units in the chain [20] (Figure 5.8). This latent manifold is deduced assuming that the water molecules surrounding the protein chain that explores conformation space are treated implicitly, that is, their influence is energetically subsumed in the potential energy function U only dependent on distances between the protein chain subunits and the local configurations that result thereof.

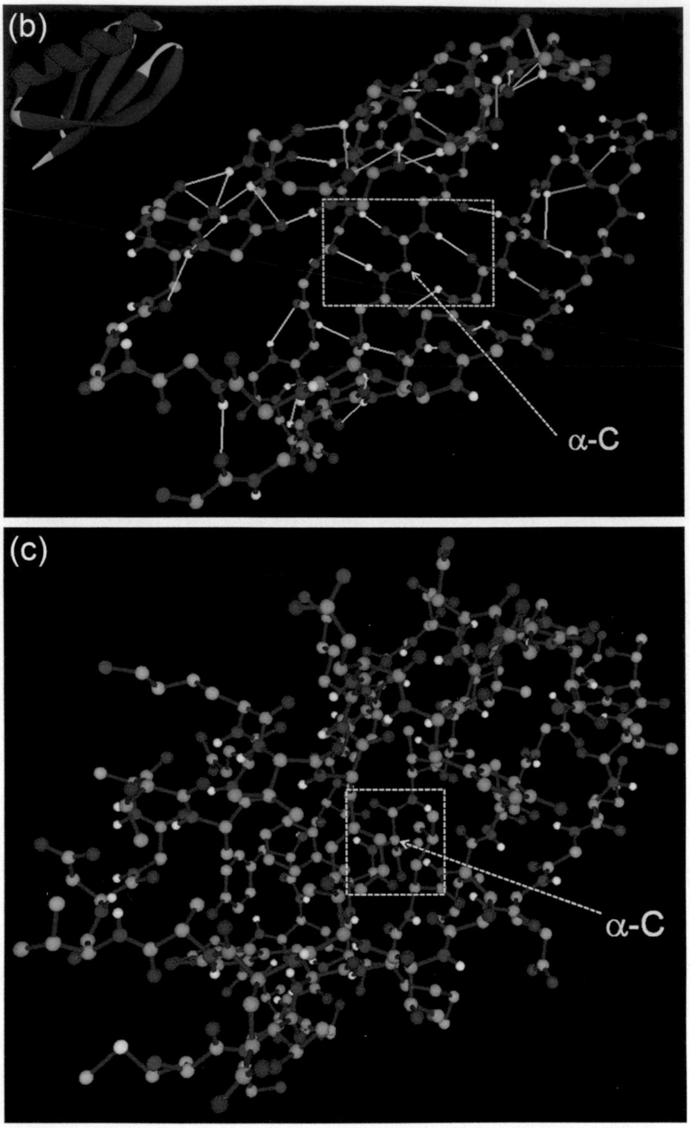

FIGURE 5.9 (b) Zoom out of the detail shown in (a), revealing the backbone HB pattern and topology (ribbon rendering) of the native structure. Backbone HBs exposed to water are indicated by thin green lines (buried ones in gray), and they signal structural deficiencies since the structure may get locally disrupted when the polar groups paired by intramolecular HBs get hydrated, that is, interact with surrounding molecules of the aqueous solvent. (c) All-atom rendering of the protein folded into its native structure as reported in PDB.1GB4.

The dynamics in W obey the basic physical law

$$\dot{z} = v; \quad \dot{v} = -\nabla_z U \tag{5.11}$$

To properly define $U=U(z)$ we introduce the following.

Theorem 5.1. There exists a map $U : \mathbb{R}^{3N}/E^{+}(3) \to \mathbb{R}$ that makes the following diagram commutative:

$$\begin{array}{ccc} \mathbb{R}^{3N} & \overset{\mathcal{U}}{\rightarrow} & \mathbb{R} \\ \pi\downarrow & \nearrow \mathrm{U} & \\ \mathbb{R}^{3N}/E^{+}(3) & & \end{array} \quad \Rightarrow \quad \mathrm{U}\circ\pi = \mathcal{U} \tag{5.12}$$

where $\pi : \mathbb{R}^{3N} \to \mathbb{R}^{3N}/E^{+}(3)$ is the canonical projection associating each point in $\mathbb{R}^{3N}$, specifying the 3D-coordinates of each of the N particles of the system, with the set of points in $\mathbb{R}^{3N}$ resulting from rotations and translations of the N-particle system treated as a rigid body, which is preserving all inter-particle distances. Thus the projection associates a state of the system with its class in quotient space, that is, with the collection of all points contained in the group orbit generated by the action of the Euclidean group on the point in the domain.

To prove this "factorization" result, it suffices to note that the potential energy $\mathcal{U} : \mathbb{R}^{3N} \to \mathbb{R}, \mathcal{U} = \mathcal{U}(\hat{z}), \hat{z} \in \mathbb{R}^{3N}$ is invariant on the orbits of $E^{+}(3)$ in $\mathbb{R}^{3N}$.

An autoencoder (γ,μ) yielding a parsimonious model in Ω would require that the autoencoder constructs $\tilde{U}(\boldsymbol{x}) = U(\mu(\boldsymbol{x}))$, so the dynamic equations in the latent manifold become

$$\dot{\boldsymbol{x}} = \tilde{\boldsymbol{v}}; \quad \dot{\tilde{v}} = -\nabla_x \tilde{\mathrm{U}}(\boldsymbol{x}). \tag{5.13}$$

This implies that in the infinitesimal limit $\tau = dt$, the potential energy function may be obtained by noting that the sparse map $\boldsymbol{K}$ defined by the autoencoder obeys

$$\dot{\boldsymbol{x}}(t) = \boldsymbol{K}(\boldsymbol{x}(t)) = -\int_0^t \nabla_x \tilde{\mathrm{U}}(\boldsymbol{x}(t'))dt' + \dot{\boldsymbol{x}}(0). \tag{5.14}$$

On the other hand, if the time series used to train the encoder is generated using molecular dynamics governed by potential energy function $U = U(\mathbf{z})$, then an additional loss term in $\dot{\boldsymbol{x}}$ of the form

$$\mathfrak{L}_U(\mu, \boldsymbol{K}) = \left\| \boldsymbol{K}(\boldsymbol{x}(t)) - \nabla_z \boldsymbol{x}(t) \left[-\int_0^t \nabla_z \mathrm{U}(z(t'))dt' + \dot{z}(0) \right]_{z(t') = \mu(\boldsymbol{x}(t'))} \right\|^2 \tag{5.15}$$

needs to be incorporated to optimize the autoencoder and its associated propagator $\boldsymbol{K}$.

5.10 METAMODELS ON LATENT MANIFOLDS

The sheer complexity of the molecular dynamics arising in soft and biological matter, especially in *in vivo* settings where $N \sim 10^6 – 10^7$ including solvent molecules, is unlikely to ultimately allow for the type of reductive approach that standard autoencoders usually provide (Figure 5.9). A case in point is the discovery of the physical underpinnings of protein folding assisted by an *in vivo* context that enhances the expediency of the process (Figure 5.10) [29]. The space is often anisotropic, the system itself is highly heterogeneous, and its components are too diverse, with potentials or force fields that cannot fully account for the complexities and many-body effects enshrined in the time series. It is doubtful that the

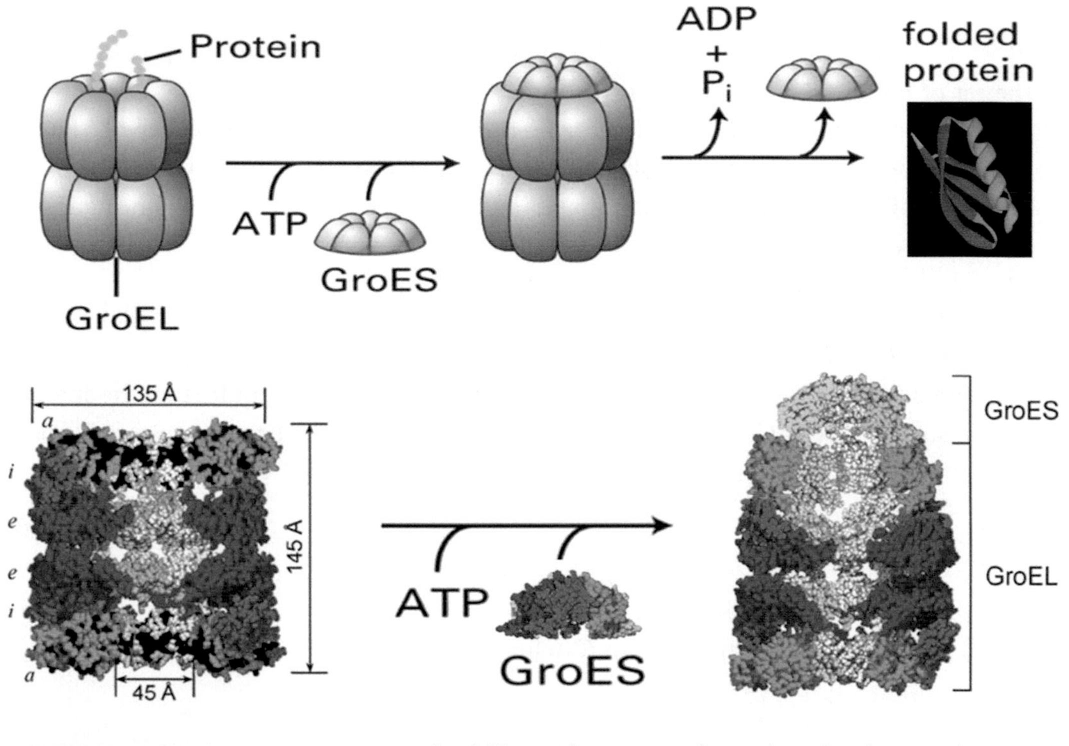

FIGURE 5.10 *In vivo* setting assisting the folding of a protein chain. A molecular cage known as chaperone GroEL assists the folding process, enhancing its expediency well above the level of efficiency that may be achieved *in vitro*, i.e., in the test tube (cf. Appendix). The cage consists of a dimer of two annular molecular assemblages, each consisting of a complex of seven identical proteins. Each protein is made of three regions known as the apical, intermediate, and equatorial domain, denoted respectively "a, i, e" in the figure. These domains undergo a certain amount of conformational rearrangement upon binding to the cell-fuel molecule ATP (adenine triphosphate). This rearrangement, in turn, enables the cage dimer to incorporate a molecular lid, known as GroES. Once the lid is on, the protein inside the cage is subject to a number of iterative annealing steps through interactions with the proteins lining the interior of the cage. This annealing process enables the protein to avoid getting kinetically trapped in misfolded states, as shown in the Appendix. Upon release of ATP, the lid becomes detached, and the protein exits the system regardless of whether it has satisfactorily completed the folding process or not. If the latter is the case, additional catalytic cycles engaging the same or other folding assistants may be necessary.

latent compact manifold spanning soft-mode coordinates (backbone torsional dihedrals in the case of the folding protein, see Figure 5.9) will be amenable to the type of model discovery that is usually cast in terms of sparse differential equations. For example, generating a minimal set of $2M$ (M~100) coupled differential equations that govern the backbone torsional dynamics underlying the protein folding process with an implicit treatment of the environment is out of reach given current capabilities in deep learning.

Other many-body systems share similar problems, as their wanton complexity is off-limits for state-of-the-art autoencoders seeking to identify parsimonious models with differential equations. Yet, as we shall now show, a generic topological understanding of the latent dynamics may yield a way of learning dynamic data that enables suitable propagation of the time series into the future, endowing the autoencoder with predictive value. Thus, topological methods will be readily incorporated in AI-based model discovery for systems with multi-scale complexity.

The approach entails a simplification based on the topological dynamics in the latent manifold, that is, on the dynamics modulo the basins of attraction of the generic singular points of the map $\boldsymbol{K}: \Omega \to T\Omega$, where $T\Omega$ denotes the tangent bundle of Ω, and Ω itself is assumed to be C^1-differentiable and compact [30], as is the case in the previously discussed example.

To make further progress with the argument, we first prove the following result.

Theorem 5.2. Under the assumptions of Section 2.3, the autoencoded map K yields no closed orbits in Ω.

By *reductio ad absurdum*, let us assume $x(0)=x(T)$ for $T>0$. Then, we get

$$0=\int_0^T \tilde{U}\left(\boldsymbol{x}(t)\right)dt=\oint_0^T d\left[\int_0^t \tilde{U}\left(x(t')\right)dt'\right]=$$

$$=\int_0^T\left[\nabla_x\int_0^t \tilde{\mathrm{U}}\left(\boldsymbol{x}(t')\right)\mathrm{d}t'\right]\dot{\boldsymbol{x}}(t)dt=\int_0^T\left[\int_0^t \nabla_x\tilde{\mathrm{U}}\left(\boldsymbol{x}(t')\right)\mathrm{d}t'\right]\dot{\boldsymbol{x}}(t)dt$$

$$=-\int_0^T \|\tilde{\boldsymbol{v}}\|^2\, dt \leq 0 \tag{5.16}$$

Equation 5.16 implies that $\tilde{\boldsymbol{v}} \equiv 0$, which is absurd since $x(0)$ was not assumed to be a steady state but a point in a closed orbit. Q.E.D.

Given that Ω is a compact manifold, this result has far-reaching consequences [30, 31]:

1) The latent dynamics have no attractors made up of recurrent orbits.
2) Since there is no circulation around them, all singular points of the latent flow are hyperbolic, hence generic, since the real part of the eigenvalues of the Jacobian at the singular points cannot be zero.

This implies the following result.

> *Corollary 5.1.* The latent dynamics governed by Equation (5.13) are of the Morse-Smale type [31], hence structurally stable, that is, qualitatively (topologically) invariant under small perturbations.

To rigorously define structural stability; we first note that the space $\mathfrak{H}(\Omega)$ of smooth maps $\Omega \rightarrow T\Omega$ is endowed with a natural metric inherited from the supremum norm given by

$$\begin{aligned} \| H \|_{sup} &= Sup_{x\in\Omega, j=1,\ldots,dim\Omega}\left[\left|\boldsymbol{H}(x)\right|, \left|\frac{\partial}{\partial x_j}\boldsymbol{H}(x)\right|\right] \\ &= Max_{x\in\Omega, j=1,\ldots,dim\Omega}\left[\left|\boldsymbol{H}(x)\right|, \left|\frac{\partial}{\partial x_j}\boldsymbol{H}(x)\right|\right] \end{aligned} \tag{5.17}$$

for $H \in \mathfrak{H}(\Omega)$. Then, to state that the latent flow $\boldsymbol{K}(\boldsymbol{x})$ is structurally stable means that for any given Δ-neighborhood of K, $\mathfrak{B}_\Delta(\boldsymbol{K}) \subset \mathfrak{H}(\Omega)$, there exists a value $\varepsilon=\varepsilon(\Delta)$, such that for any $\boldsymbol{G} \in \mathfrak{B}_\Delta(\boldsymbol{K})$ there exists an ε-homeomorphism $h_\varepsilon : \Omega \rightarrow \Omega$ satisfying $\max_{x\in\Omega} |\boldsymbol{x} - h_\varepsilon(\boldsymbol{x})| < \varepsilon$ that transforms trajectories of $\boldsymbol{K}$ onto trajectories of $\boldsymbol{G}$ [31].

Given the qualitative invariance of the latent flow under small perturbations, the following observation is key to justify the leverage of AI to construct dynamic models based on time series: *The structural stability of the latent dynamics is essential to enable model discovery in view of the fact that the exact parameters determining the potential energy U of the many-body system are not known precisely.*

Given the characterization of the latent flow given by Theorem 5.2 and Corollary 5.1, we may encode the flow in a simplified manner, as we now build a metamodel. To that effect, we first define the equivalence relation "~" for any pair $\boldsymbol{x}, \boldsymbol{y} \in \Omega : \boldsymbol{x} \sim \boldsymbol{y} \Leftrightarrow \omega(\boldsymbol{x}) = \omega(\boldsymbol{y})$, where $\boldsymbol{\omega}(x)$ denotes the destiny (omega) state given by the limit at $t\rightarrow\infty$ of the trajectory initiated at $\boldsymbol{x}$ [32]. Since the singular points are the steady states of the system, the latent flow may be encoded by the equivalence classes identified as the basins of attraction of the singular points. As points in Ω are regarded "modulo basins," we have in effect defined the quotient space $\Omega/\sim$ as the set of basins of attraction and separatrices (basins of lower dimension) of critical generic points that partition the manifold. The quotient space is relatively simple to encode since the singular points of the latent flow are isolated and finite in number. To demonstrate this proposition, we note that otherwise, if they were infinite in number, they would have an accumulation point since the latent manifold is compact, and that cannot happen because all singular points are generic.

Let us then denote by Γ: $\Omega/\sim \rightarrow \Omega/\sim$ the coarse-grained flow that determines the inter-basin transitions, while π: $\Omega \rightarrow \Omega/\sim$denotes the canonical projection that associates a point in the latent manifold with its equivalence class. The flow must be such that the diagram in Figure 5.11 becomes commutative, and specifically, the following flow-compatibility relations must hold:

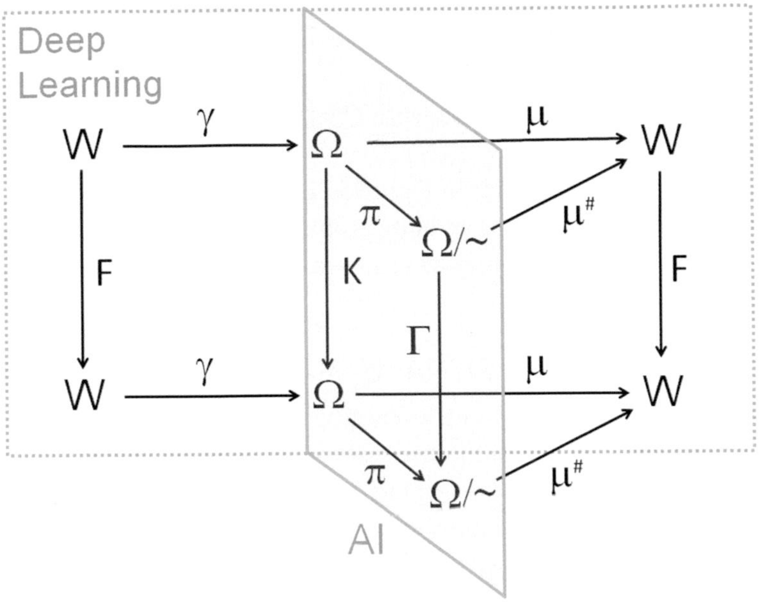

FIGURE 5.11 Scheme of a metamodel consisting of two coupled variational autoencoders $(\gamma,\mu),(\pi,\mu^{\#})$ required to generate the discrete flow Γ that propagates the topological dynamics in the latent quotient manifold Ω/~. The parameter optimization of both autoencoders ensures the full commutativity of the diagram. The discovery of the topological metamodel $\Gamma: \Omega/\sim \rightarrow \Omega/\sim$ hinges on a CNN-based construction of the modulo-basin projection $\pi: \Omega \rightarrow \Omega/\sim$ of the latent dynamics. This scheme introduces a level of coarse-graining that is more drastic than that adopted by conventional autoencoders for model discovery of time series data.

$$\mu^{\#} \circ \pi = \mu \tag{5.18}$$

$$\Gamma \circ \pi = \pi \circ K \tag{5.19}$$

$$\mu^{\#} \circ \Gamma = F \circ \mu^{\#} \tag{5.20}$$

Thus, to parametrize Γ, we need to introduce a second autoencoder with variational parameter optimization determined by the loss function

$$\mathfrak{L}_{\sim}\left(\mu^{\#},\Gamma\right) = \mathfrak{L}_{\mu}\left(\mu^{\#}\right) + \mathfrak{L}_{\sim}\left(\Gamma\right) + \mathfrak{L}_{\sim}\left(\mu^{\#}\right), \tag{5.21}$$

where:

$$\mathfrak{L}_{\mu}\left(\mu^{\#}\right) = Q^{-1} \sum_{x \in \mathfrak{F}} \| \left(\mu^{\#} \circ \pi - \mu\right) \boldsymbol{x} \|^2, \tag{5.22}$$

$$\mathfrak{L}_{\sim}\left(\Gamma\right) = Q^{-1} \sum_{x \in \mathfrak{F}} \| \left(\Gamma \circ \pi - \pi \circ K\right) \boldsymbol{x} \|^2, \tag{5.23}$$

$$\mathfrak{L}_{\sim}\left(\mu^{\#}\right)=Q^{-1}\sum_{x\in\mathfrak{F}}\|\left(\mu^{\#}\circ\Gamma-F\circ\mu^{\#}\right)\pi x\|^{2}, \tag{5.24}$$

where the training set is denoted $\mathfrak{F}\subset\Omega$, with $Q=|\mathfrak{F}|$ and $\mu^{\#}$ denoting the decoder for the quotient space.

The commutativity of the diagram in Figure 5.11 implies that the ultimate simplicity in a model governing ultra-complex many-body problems, such as identifying *in vivo* protein folding trajectories (cf. Appendix), may be achieved by projecting the latent dynamics onto the quotient manifold $\Omega/\sim$.

5.11 QUOTIENT SPACES FOR DYNAMICAL SYSTEMS

The sparse latent dynamics obtained by leveraging autoencoders that serve as model discoverers have become the subject of intense research in applied mathematics. Such methods are less suited to unraveling underlying laws in dynamical systems that represent biological or soft matter, where the number of internal degrees of freedom is astronomical. This book proposes to couple two commutative – hence compatible – autoencoders as described in Figure 5.12, yielding a factorization of the latent dynamics through the quotient space $\Omega/\sim$. The commutativity of the whole diagram displayed in Figure 5.12 ensures the compatibility of the different levels of coarse-graining of the dynamics that constitute the metamodel. In turn, the diagram commutativity (Equations (5.19) and (5.20)) is

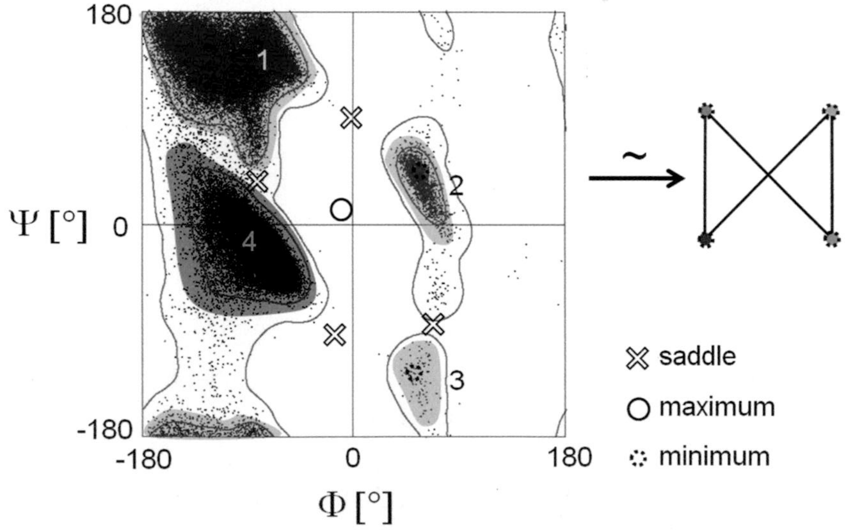

FIGURE 5.12 Toroidal (Φ, Ψ) cross-section of the time series data for an individual unit along a folding protein chain. The associated cross-section of the dynamical system representing the protein folding process is discovered through pattern recognition leveraging a CNN, and its modulo-basin representation in the cross-section of the latent quotient manifold is given as a graph. The vertices in the graph are critical points corresponding to omega-sets for all points in their respective basins of attraction, while the edges indicate allowed inter-basin transitions that determine the topological dynamics.

subsumed into the variational functional of the autoencoder (Equations (5.21)–(5.24)) so that optimization of the underlying neural networks is equated with – or rather becomes as close as possible to – diagram commutativity. Identifying the quotient space $\Omega/\sim$ under the "modulo-basin" equivalence relation defined in the previous section is akin to a pattern recognition process, where time series datapoints are plotted onto 2D cross-sections of the latent manifold Ω. The task may be entrusted to a CNN and becomes enormously simplified as Corollary 5.1, jointly with Theorem 5.2, and guarantees a finite number of basins of attraction for isolated singular points which are all generic, that is, topologically equivalent under small perturbations of the latent vector field.

For example, in the case of the folding protein, the 2D cross-section is the (Φ,Ψ)-torus (Figure 5.9a), and a typical time series (details in Appendix) for latent dynamics governed by Equation 5.13 is given in Figure 5.12 [19, 20]. The CNN "discovers" the topological metamodel which can be represented as a graph with vertices corresponding to the basins of attraction of the generic minima and edges connecting basins of attraction (Figure 5.12) in a manner specified by the autoencoder-generated information fed onto the CNN. More specifically, the graph is generated in accord with the inferred topography of the 2-torus cross-section of the potential energy map $\tilde{U}:\Omega \to \mathbb{R}$, in turn generated by the autoencoder.

To generate the entire quotient space within a graph representation, we need to integrate the cross-sections corresponding to the 2D projections of $\Omega/\sim$. As an illustration, let us consider a GG dipeptide (M=2, G=glycine). The latent manifold is a Cartesian product of two tori (Figure 5.13), one for each pair of Φ, Ψ dihedral coordinates specifying the

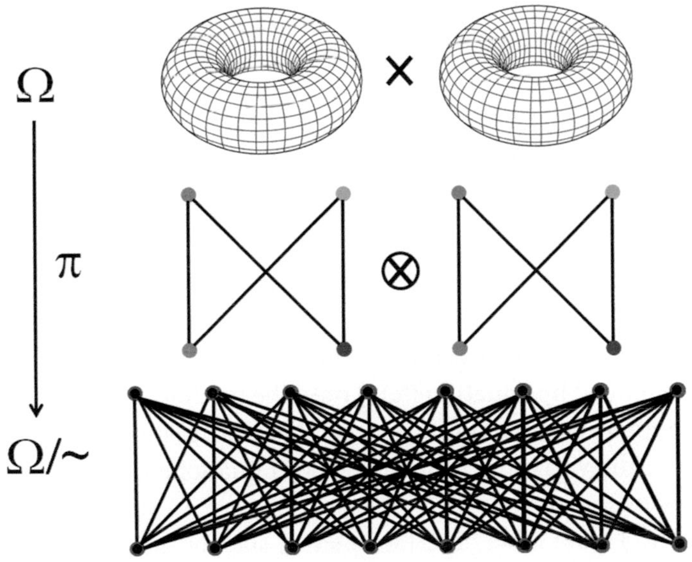

FIGURE 5.13 Reconstruction of the latent quotient manifold. The iterative process is represented by progressively incorporating adjacent-residue cross-sections of the latent manifold as Cartesian products and, in parallel, reconstructing the quotient manifold as a tensorial product of the individual graphs representing the cross-sections of the quotient manifold (Figure 5.12). Within this representational framework, the topological dynamics become a walk in the quotient manifold graph.

backbone conformation of the respective residue [20]. The topological representation of the vector field steering the backbone torsional dynamics of a generic residue in the protein chain is given in Figure 5.12. Opposite sides of the square are identified as per the ±180° identification of Φ, Ψ dihedrals determining the local torsional state of the backbone. The four colored sectors morph topologically into the allowed valleys in potential energy. The organization of the basins of critical points is compatible with the underlying 2-torus and is topologically represented by a graph (Figure 5.12) for each residue. The bottom panel in Figure 5.13 represents the quotient space for the dipeptide chain [20]. For each residue, the quotient space cross-section is represented by a graph with vertices indicating two-dimensional basins, and an edge linking two vertices indicates that the respective basins are connected through a line of steepest descent crossing a saddle point and orthogonal to the separatrix at the saddle. For a protein chain consisting of two consecutive residues, denoted 1 and 2, the quotient space becomes the tensor product of the two graphs where vertices now denote basin pairs (B_1, B_2) and where the basin pair (B_1, B_2) is connected with (B'_1, B'_2) if and only if B_1 is connected with B'_1 or $B_1=B'_1$, and B_2 is connected with B'_2 or $B_2=B'_2$. Thus, the quotient space for two residues consists of 4×4=16 vertices, where each vertex connects via one edge with eight other vertices and connects via two adjacent edges to the seven remaining vertices. Given the symmetry of the problem, to prove the assertion, it suffices to note that the vertex denoting basin pair (1,1) is directly connected to eight other vertices denoting pairs (2,1), (4,1), (1,2), (1,4), (2,2), (2,4), (4,2), (4,4) while vertex (1,1) connects to all the remaining seven vertices-pairs containing basin 3 via two adjacent edges.

Thus, the projection on quotient space $\Omega/\sim$ of a latent MD trajectory on the 4-torus Ω becomes a walk on the tensorial product graph at the bottom of Figure 5.13 [20].

5.12 MOLECULAR DYNAMICS ENCODED AND PROPAGATED AS TOPOLOGICAL DYNAMICS

The coupling of two autoencoders where all flows commute as described in Figure 5.11 becomes essential to discover topological metamodels governing the ultra-complex dynamics encountered in a biological matter. Such systems are unlikely to yield sparse differential equations governing the latent dynamics in a compact manifold, but that is only an educated guess. To train both autoencoders in parallel, it is necessary to project the latent dynamics generated by Equation (5.13) onto the modulo-basin topological dynamics defined on the quotient space and use the encoded information to optimize the autoencoders according to the variational functionals of Equations (5.5)–(5.9) and (5.21)–(5.24) defined in the previous sections.

As an illustration, Figure 5.14a describes a latent state of the protein chain with chemical composition (amino acid sequence) corresponding to human *ubiquitin* [19]. The state is represented as a point in the latent compact manifold $\Omega=\prod_{n=1}^{M}\Omega_n$ (M=76), where Ω_n is the two-torus spanned by the two backbone torsional dihedral coordinates of the nth

	1	2	3	4	5	6	7	8	9	10
Aminoacid	MET-M	GLN-Q	ILE-I	PHE-F	VAL-V	LYS-K	THR-T	LEU-L	THR-T	GLY-G
Phi-angle	56.50	-92.00	-129.00	-111.00	-115.00	-89.00	-95.00	-73.00	-94.00	88.00
Psi-angle	34.92	133.00	160.00	134.00	106.00	121.00	170.00	-11.00	3.00	14.00
Omega-angle	180.00	180.00	180.00	180.00	180.00	180.00	180.00	180.00	180.00	180.00
	11	12	13	14	15	16	17	18	19	20
Aminoacid	LYS-K	THR-T	ILE-I	THR-T	LEU-L	GLU-E	VAL-V	GLU-E	PRO-P	SER-S
Phi-angle	-102.00	-112.00	-108.00	-96.00	-128.00	-107.00	-137.00	-116.00	-55.00	-83.00
Psi-angle	131.00	128.00	136.00	138.00	150.00	114.00	168.00	142.00	-16.00	-45.00
Omega-angle	180.00	180.00	180.00	180.00	180.00	180.00	180.00	180.00	180.00	180.00
	21	22	23	24	25	26	27	28	29	30
Aminoacid	ASP-D	THR-T	ILE-I	GLU-E	ASN-N	VAL-V	LYS-K	ALA-A	LYS-K	ILE-I
Phi-angle	-74.00	-81.00	-64.00	-57.00	-71.00	-62.00	-61.00	-68.00	-64.00	-64.00
Psi-angle	143.00	161.00	-37.00	-39.00	-37.00	-43.00	-34.00	-37.00	-42.00	-39.00
Omega-angle	180.00	180.00	180.00	180.00	180.00	180.00	180.00	180.00	180.00	180.00
	31	32	33	34	35	36	37	38	39	40
Aminoacid	GLN-Q	ASP-D	LYS-K	GLU-E	GLY-G	ILE-I	PRO-P	PRO-P	ASP-D	GLN-Q
Phi-angle	-64.00	-58.00	-94.00	-118.00	84.00	-79.00	-58.00	-56.00	-64.00	-92.00
Psi-angle	-43.00	-38.00	-27.00	-11.00	5.00	119.00	136.00	-35.00	-18.00	-10.00
Omega-angle	180.00	180.00	180.00	180.00	180.00	180.00	180.00	180.00	180.00	180.00
	41	42	43	44	45	46	47	48	49	50
Aminoacid	GLN-Q	ARG-R	LEU-L	ILE-I	PHE-F	ALA-A	GLY-G	LYS-K	GLN-Q	LEU-L
Phi-angle	-85.00	-128.00	-95.00	-122.00	-143.00	51.00	72.00	-112.00	-84.00	-75.00
Psi-angle	130.00	107.00	131.00	130.00	128.00	43.69	-21.00	140.00	122.00	134.00
Omega-angle	180.00	180.00	180.00	180.00	180.00	180.00	180.00	180.00	180.00	180.00
	51	52	53	54	55	56	57	58	59	60
Aminoacid	GLU-E	ASP-D	GLY-G	ARG-R	THR-T	LEU-L	SER-S	ASP-D	TYR-Y	ASN-N
Phi-angle	-95.00	-45.00	-75.00	-82.00	-103.00	-61.00	-66.00	-58.00	-97.00	63.00
Psi-angle	133.00	-40.00	-20.00	164.00	165.00	-33.00	-31.00	-32.00	-1.00	34.00
Omega-angle	180.00	180.00	180.00	180.00	180.00	180.00	180.00	180.00	180.00	180.00
	61	62	63	64	65	66	67	68	69	70
Aminoacid	ILE-I	GLN-Q	LYS-K	GLU-E	SER-S	THR-T	LEU-L	HIS-H	LEU-L	VAL-V
Phi-angle	-77.00	-93.00	-53.00	73.00	-74.00	-117.00	-97.00	-105.00	-105.00	-112.00
Psi-angle	112.00	169.00	140.00	18.00	157.00	122.00	151.00	129.00	117.00	133.00
Omega-angle	180.00	180.00	180.00	180.00	180.00	180.00	180.00	180.00	180.00	180.00
	71	72	73	74	75	76				
Aminoacid	LEU-L	ARG-R	LEU-L	ARG-R	GLY-G	GLY-G				
Phi-angle	-91.00	-122.00	-111.00	-111.00	-56.99	-64.99				
Psi-angle	138.00	95.00	120.00	120.00	-46.01	-59.81				
Omega-angle	180.00	180.00	180.00	180.00	180.00	180.00				

FIGURE 5.14 Latent manifold and latent quotient manifold for a folding protein. (a) Dihedral torsional representation of a point in the latent manifold for the folding of ubiquitin ($M=76$, $\dim\Omega=152$). (b) Equivalence class in the latent quotient manifold containing the point given in (a). The four basins of attraction are represented by colored quadrants in a square.

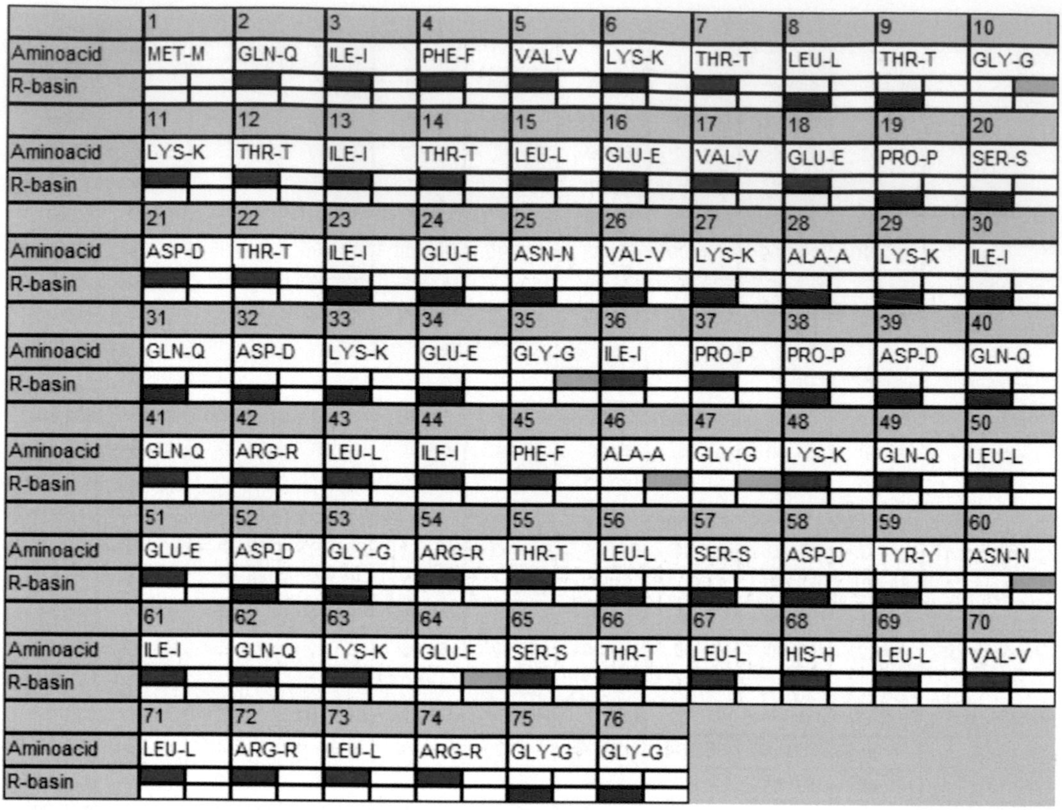
1 2 3 4 5 6 7 8 9 10
Aminoacid MET-M GLN-Q ILE-I PHE-F VAL-V LYS-K THR-T LEU-L THR-T GLY-G
R-basin
11 12 13 14 15 16 17 18 19 20
Aminoacid LYS-K THR-T ILE-I THR-T LEU-L GLU-E VAL-V GLU-E PRO-P SER-S
R-basin
21 22 23 24 25 26 27 28 29 30
Aminoacid ASP-D THR-T ILE-I GLU-E ASN-N VAL-V LYS-K ALA-A LYS-K ILE-I
R-basin
31 32 33 34 35 36 37 38 39 40
Aminoacid GLN-Q ASP-D LYS-K GLU-E GLY-G ILE-I PRO-P PRO-P ASP-D GLN-Q
R-basin
41 42 43 44 45 46 47 48 49 50
Aminoacid GLN-Q ARG-R LEU-L ILE-I PHE-F ALA-A GLY-G LYS-K GLN-Q LEU-L
R-basin
51 52 53 54 55 56 57 58 59 60
Aminoacid GLU-E ASP-D GLY-G ARG-R THR-T LEU-L SER-S ASP-D TYR-Y ASN-N
R-basin
61 62 63 64 65 66 67 68 69 70
Aminoacid ILE-I GLN-Q LYS-K GLU-E SER-S THR-T LEU-L HIS-H LEU-L VAL-V
R-basin
71 72 73 74 75 76
Aminoacid LEU-L ARG-R LEU-L ARG-R GLY-G GLY-G
R-basin

FIGURE 5.14 Continued

residue along the chain. Using the modulo-basin encoding π: $\Omega \rightarrow \Omega/\sim$ defined in Figure 5.13, the state given in Figure 5.14a becomes part of the equivalence class in $\Omega/\sim$ defined by the modulo-basin topology given in Figure 5.14b. This can be verified residue by residue using Figure 5.12. Once the two coupled autoencoders are trained in parallel, the system learns to propagate dynamics in the latent quotient space $\Omega/\sim$, as described in Figure 5.15. The training process, in this case, involved a laborious 0.5ms-MD simulation of ubiquitin folding assisted by the chaperone GroEL (Figure 5.10), as detailed in the Appendix. The decoding $\mu^{\#}$: $\Omega/\sim \rightarrow W$ of the modulo-basin topology generated by the learned flow Γ at 7ms yields a protein structure (Figure 5.15), which is very close (RMSD=1.15Å) to the native fold reported in the entry 1UBI of the protein data bank (PDB). This experimental validation attests to the power of leveraging topological dynamics for AI-enabled model discovery of biological matter.

5.13 AUTOENCODER BATTERIES FOR HIERARCHICAL SYSTEMS

Kurzweil and others have successfully argued that a hierarchical structure of reality is necessary for proper encoding and processing in a suitable AI-based inferential framework [22, 33]. To cast the discussion in the broadest terms, we shall refer to hierarchical as the attribute of a system endowed with nested complexities at multiple scales arising

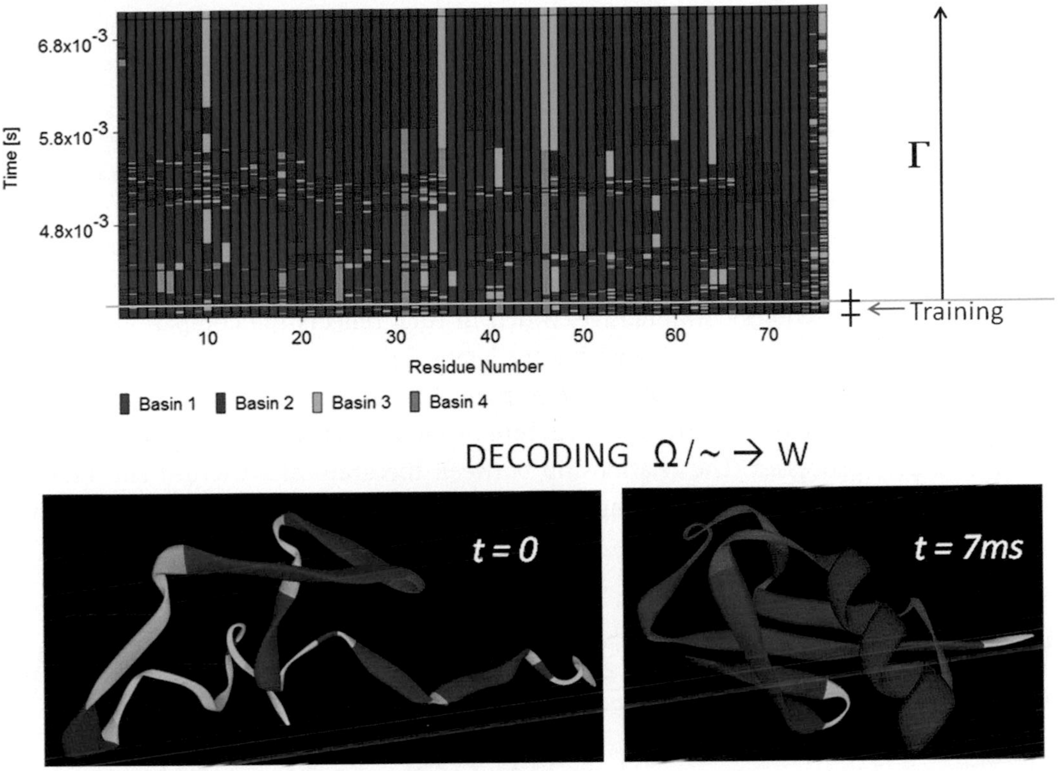

FIGURE 5.15 Latent folding dynamics generated by two coupled autoencoders (Figure 5.11) yielding an *in-vivo*-assisted pathway that expeditiously yields the native fold of ubiquitin (PDB.1UBI). The autoencoders are trained by time series obtained by running MD trajectories spanning 0.5ms of a molecular *in vivo* setting (chaperone, Figure 5.10) that steers the folding chain. Each horizontal line in the plot represents a modulo-basin conformation of the entire chain resolved at time intervals of 50μs. Hence, the plot describes the AI-empowered propagation of the folding dynamics projected on the latent quotient space. The plot constitutes a topological dynamics metamodel of the folding of ubiquitin.

at different levels of description and providing different levels of coarse-graining. This notion was delineated previously in this chapter, where a number of illustrations reveal the dynamic entrainment of fast-relaxing modes by slower modes spanning a latent manifold. Thus, just like in the adiabatic approximation [16], fast motions are averaged out and hence treated implicitly in a coarse-grained version of the dynamics focusing on longer timescales, with an autoencoder providing the inferential framework to discover the latent coordinate system.

Indeed, an escalation in the level of coarse-graining, from the subatomic to the atomic, molecular, subcellular, and beyond, has always suggested that the hierarchical structure of reality may span over several layers, with nested complexities where information at a peripheral layer is incorporated implicitly at a core layer. In topological terms, we envision a whole sequence of quotient manifolds, as we lump up states at different levels of description within progressively coarser equivalence classes, with dynamic compatibility of the various descriptions imposed by commutative flow diagrams (cf. Figure 5.16). Thus, in the broadest sense, an encoding of a hierarchical dynamical system within an AI-based inferential framework may require a battery of tandem autoencoders, whose dynamic compatibility in an optimized parametrization is ensured by the commutativity of the diagram that combines the different levels of flow encoding. Figure 5.16 shows one such diagram reflecting the dynamic interplay of three autoencoders, one to generate the latent manifold Ω of intrinsic coordinates, one for a first-level quotient space $\Omega/\sim$, and one for a second-level (coarser) quotient space $\Omega/\sim/\approx$. The autoencoders are variationally optimized so that the flow diagram becomes commutative, which in turn reflects the compatibility of the different levels of dynamic encoding within the latent three-level hierarchy.

The subsequent chapters illustrate the implementation of tandem autoencoder technology to treat implicitly the hierarchical structure of the molecular dynamics of biological and biomedical processes. The relationship between hierarchical structure and tandem autoencoders of nested coarse-graining descriptions is of universal applicability and may

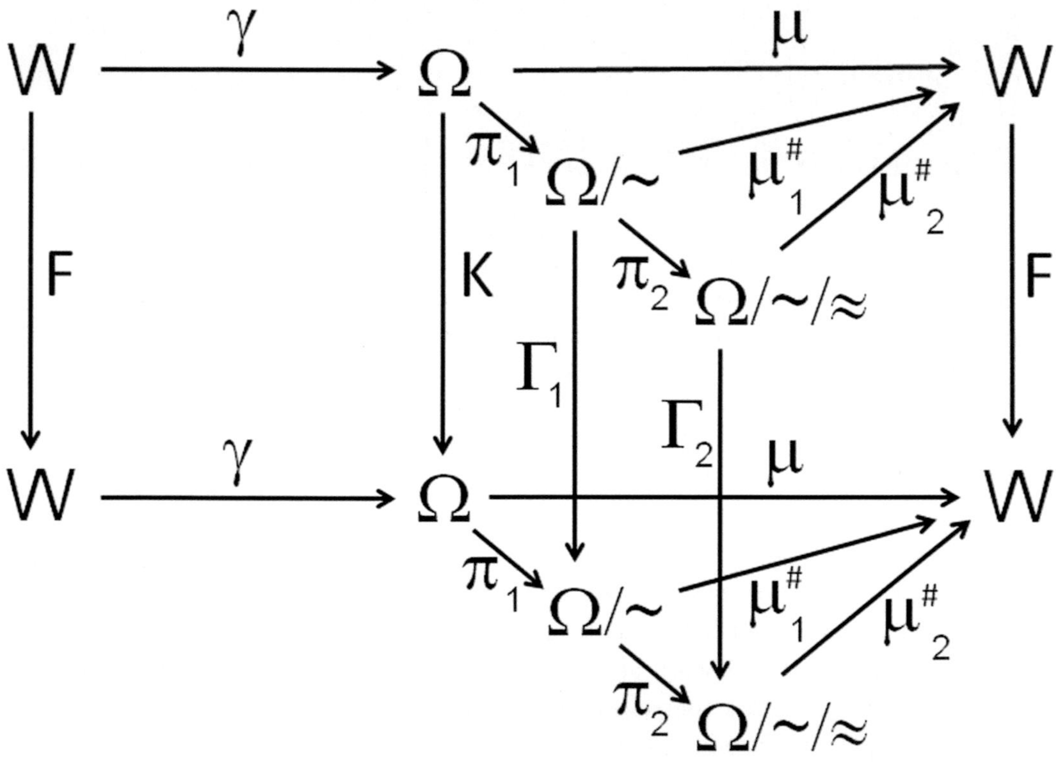

FIGURE 5.16 Metamodels for hierarchical dynamical systems. Autoencoders in tandem implemented to forecast dynamics that allow for a hierarchical description defined by two equivalence relations "$\sim$" and "$\approx$." Two autoencoders $\{\pi_1, \mu^{\#}_1, \Gamma_1\}$ and $\{\pi_2, \mu^{\#}_2, \Gamma_2\}$ operating sequentially as required for topological encoding of the dynamics at the coarsest level in the latent quotient manifold $\Omega/\sim/\approx$. The F-compatibility of the latent flows K, Γ_1, Γ_2 in Ω, $\Omega/\sim$, $\Omega/\sim/\approx$, respectively, requires the full commutativity of the diagram. This commutativity is achieved by variational parameter optimization for the three autoencoders $\{\gamma, \mu, K\}$, $\{\pi_1, \mu^{\#}_1, \Gamma_1\}$, and $\{\pi_2, \mu^{\#}_2, \Gamma_2\}$.

be extended to any number of equivalence relations, leading to progressively simplified metamodels.

5.14 DECODING QUANTUM GRAVITY

A theory of quantum gravity (QG), that is, a theory of gravity in accord with the tenets of quantum mechanics (QM), is often regarded as off-limits due to the major disparity in the dimensional scales where quantum and gravitational forces materialize [34, 35]. So far, gravity has not been quantized and hence a sense of incompleteness pervades the field. The unification of all forces of nature remains a holy grail in physics ever since Einstein's pursuit.

A different sort of grand unification was pursued by Albert Einstein. Since weak and strong nuclear forces were not clearly understood at the time, he sought to do it without involving QM, a theory he never fully endorsed. At first glance, quantum gravity stands almost as an oxymoron: After all, QM deals with the atomic and subatomic scales, while the best theory of gravity to date is Einstein's general relativity (GR), which is essentially classic, i.e., non-quantum, and deals mainly with cosmological scales (except for singularities). Einstein's theory of relativity postulates that high concentrations of energy and matter impinge on the curvature of space-time, deflecting the trajectories of particles, as it occurs in a gravitational field. This theory withstood admirably the long-term attrition of experimental corroboration. Yet, if we attempt to cast GR in QM terms, we need to deal with the fact that matter and space-time become "protean" at scales of the order of Planck's length (10^{-33}cm), akin to the sea of virtual particles that fill up empty space. In this essentially quantum world, the equations of GR no longer hold.

Pursuing QG still makes good sense when we deal with GR singularities, for example, black holes or the first few sub-attoseconds after the Big Bang, when the vast differences in material scales that set apart QM and GR can be reconciled, prior to the inflationary phase. Thus, in the spirit of this book, we need to address the question: *What would constitute a quantum metamodel of a theoretical model of gravity, and how could this be constructed?*

This question was effectively formulated by the Argentinian physicist Juan Maldacena [34]. He focused on the anti-de Sitter (AdS) space, a hyperbolic space that shares curvature properties with the sphere representing the event horizon of a black hole. Maldacena postulated that a string theory (ST) of gravity in a five-dimensional AdS space (AdS_5=W in standard notation) is *equivalent* to a quantum field theory (QFT) on its boundary ∂W, which constitutes a four-dimensional Minkowski space, akin to the one Einstein adopted for GR.

How could the boundary represent the latent space for the ST in an AdS? The answer is provided by the computation of the Shannon entropy of the black hole that assesses its total information storage capacity contained in all degrees of freedom [36]. These "ultimate" degrees of freedom involve of course atomic and subatomic entities, all the way to quarks and gluons, and ultimately those entities from hitherto unknown depths in the physical structure of matter, the so-called "level X." This "ultimate entropy" is proportional to the surface area of the event horizon [36].

In general, the boundary M_d of AdS_{d+1} is a d-dimensional Minkowski space with the symmetry group $SO(2,d)$ of AdS_{d+1} acting on M_d as the conformal (i.e., inner product-preserving) group. Thus, there are two ways to get a physical theory with $SO(2,d)$ symmetry: A relativistic field theory on AdS_{d+1} and a conformal field theory (CFT) on M_d. A suitable theory on AdS_{d+1} has been conjectured by Maldacena to be *equivalent* to a CFT on M_d [34]. The computation of observables of the CFT in terms of supergravity on AdS_{d+1} can and should be attempted using the methods described in this book. In accordance with the tenets of topological metamodel auto-encoding, M_d should be identified with the latent manifold Ω, and the CFT on M_d "holographically" spanned onto AdS_{d+1} should be generated by a holographic autoencoder that exploits the $SO(2,d)$ to generate jointly the latent manifold together with its parsimonious metamodel.

Correlation functions in conformal field theory (QFT) are given by the dependence of the supergravity action on the asymptotic behavior at infinity [35]. Thus, the dimensions of operators in CFT are determined by masses of particles in string theory. It is thus conjectured that to describe the Yang-Mills theory in four dimensions, one should use the whole infinite tower of massive Kaluza-Klein states on $AdS_5 \times \mathbf{S}^5$. Chiral fields in the four-dimensional $\mathcal{N}=4$ theory correspond to Kaluza-Klein harmonics on $AdS_5 \times \mathbf{S}^5$. The spectrum of Kaluza-Klein excitations of $AdS_5 \times \mathbf{S}^5$ is matched against operators of the $\mathcal{N}=4$ theory.

To discover the topological metamodel for superstring theory on $W = AdS_{d+1} \times \mathbf{S}^{d+1}$, we first note that the boundary is topologically identified as $\partial W = (S^1 \times S^{d-1}) / \mathbb{Z}_2$, where the group $\mathbb{Z}_2$ acts by rotation in π on S^1 and multiplication by -1 on S^{d-1}. In other words, the latent manifold fulfills the compactness condition, and to simplify the computation and straighten the symmetry we may use its universal cover [37], as demonstrated by Ariel Fernández and Oktay Sinanoglu: $\Omega \approx S^{d-1} \times \mathbb{R}$, with the real axis representing the time dimension.

Thus, to prove using AI the conjectured equivalence between $\mathcal{N}=4$ QFT on $\Omega \approx S^{d-1} \times \mathbb{R} = \partial W$ and Type IIB supergravity as string theory (ST) on $W = \langle AdS_5 \times \mathbf{S}^5 \rangle$ ($\langle . \rangle$=universal cover [37]), we need a holographic autoencoder for a DL neural network capable of representing the ST on the space W. This DL system has been constructed [38]. The autoencoder should identify the latent manifold as $\Omega \approx S^3 \times \mathbb{R}$ taking advantage of the $SO(2,4)$ symmetry with which the AdS space is endowed. In this way, the projection π onto the latent manifold, identified by the autoencoder as ∂W, constitutes the inverse of the postulated holographic map $h{:}\partial W \rightarrow W$ that makes the following diagram commutative (Figure 5.17):

$$\begin{array}{ccc} W & \overset{\pi}{\rightarrow} & \partial W \\ \downarrow F_{ST} & & \downarrow K_{QFT} \\ W & \overset{h}{\leftarrow} & \partial W \end{array} \tag{5.25}$$

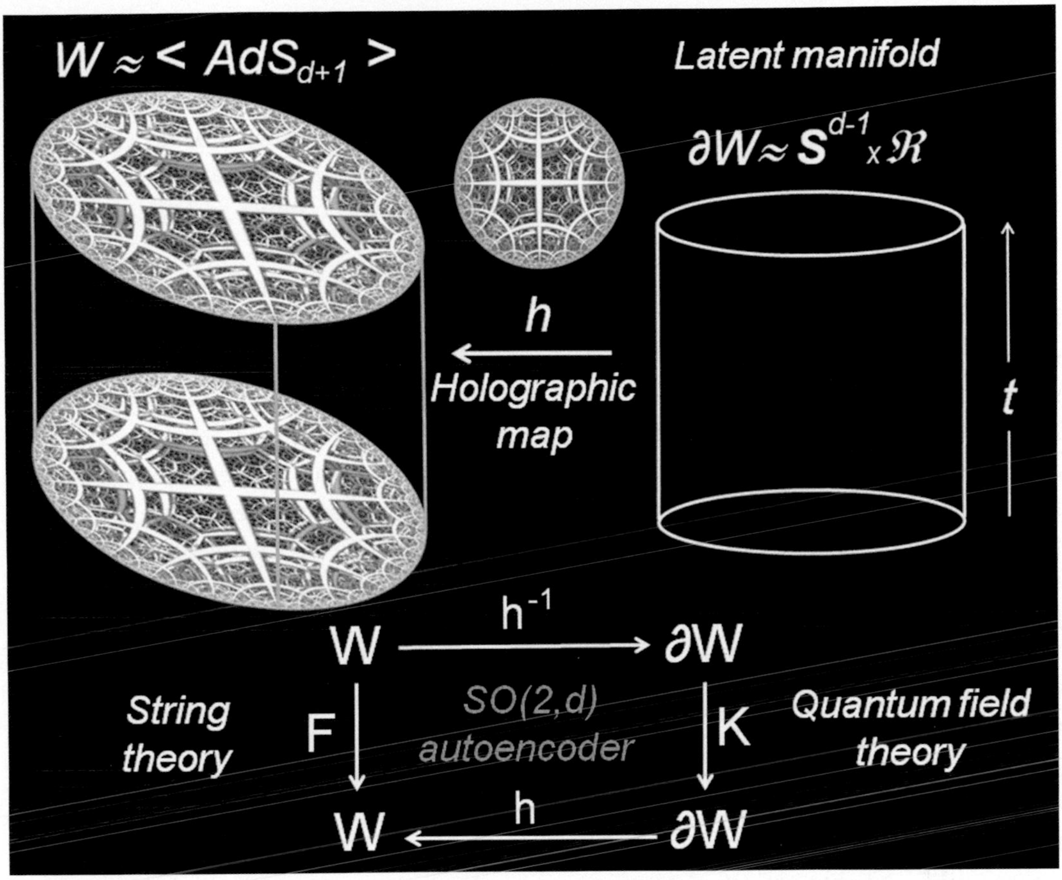

FIGURE 5.17 Holographic autoencoder enabling the discovery of a quantum metamodel of gravity.

Thus, the AdS/CFT equivalence may be proved by AI. The program entails two steps: (a) Generation of a data-driven holographic metamodel of quantum systems by formulating its supergravity dual/equivalent on a DL NN, and (b) construction of the appropriate autoencoder (π, K_{QFT}) that fits the holographic map h, yielding the functional identity $h \circ K_{QFT} \circ \pi = F_{ST}$ of homeomorphisms on W. Part a) of the program has already been achieved, as shown elsewhere [38]. A deep NN representation of the AdS/CFT correspondence has been obtained, with the emergence of the bulk metric based on deep learning of data generated as an outcome of boundary quantum field theories. The radial direction of the bulk metric is assimilated with the depth of the hidden layers. Thus, the network provides a data-driven model of strongly coupled systems. In the space-time for a black hole horizon, the deep NN can fit boundary data generated by the AdS Schwarzschild space-time, reproducing the metric as the data is reverse-engineered. With inputted experimental data, the deep NN determines the bulk metric, mass, and quadratic coupling. Thus, the NN provides a gravitational model of strongly correlated systems.

As for part (b), the holographic autoencoder (HAE) that will ultimately identify the latent manifold with ∂W can be obtained using the methods described in this chapter. The remaining challenge in part (b) is for the HAE to simultaneously yield the parsimonious QFT at the boundary ∂W that serves as a metamodel for the gravity model on W in the sense adopted in this book. To address this challenge requires that we harness the isomorphism already noted [39] between a deep autoencoder and the multi-scale entanglement renormalization ansatz (MERA).

If the universe is indeed a hologram, as Maldacena and others suggest [34], the inverse of the holographic map of its avatar, the event horizon of the black hole, represents a one-to-one (injective) projection onto the horizon boundary. Therefore this boundary may be regarded as a model for the latent quotient manifold of the universe. In this quotient manifold, one dimension has been folded up and stored within the equivalence classes that are ultimately the only observables and hence the only encoded features endowed with a physical entity. This topology for the quintessential universe is not entirely correct but locally correct, as shown in Chapter 6. The overall topology of the universe is immaterial to this discussion, as both quantum physics and GR are in fact local theories.

Ultimately, quantum gravity will be ascertained through a holographic autoencoder that identifies the "correct" latent manifold and associated quantum metamodel, or perhaps by other means available to theoretical physicists. Be that as it may, we may state that all events following the Planck epoch of the Big Bang ($t > 10^{-43}$s [40]) are likely to be of a quantum nature. This is so because, at that stage, gravity branches off from the three other fundamental forces already accounted for by quantum theory, so all four differentiated forces will be reliably identified as quantum forces. A rigorous topological characterization of the universe supports this assertion, as shown in Chapter 6, with gravity originally stored as a de Broglie wave in a compact fifth dimension. Thus, the certainty of all events that follow the Planck epoch can only be established through the participation of observers. This singular circumstance leaves us having to postulate God's existence at least as early as 10^{-43}s after the Big Bang or admit that the universe remains a mere possibility, replete with phenomena-to-be within multiple *a priori* potential realities that are equally possible in the quantum realm, as in the multiverse scenario.

5.15 QUANTUM GRAVITY AUTOENCODER

To address the problem of quantum gravity from an AI-borne perspective, we need to design a holographic autoencoder by leveraging the physics of machine learning to address the problem of whether emergent quantum behavior can arise in a neural network. By emergent quantum mechanics, we mean a formulation within a framework of nonlocal hidden variables, as in the Bohm scheme [41]. Once emergent quantum behavior is shown to become possible in machine learning physics, we may address the question of developing a relativistic string gravitational scheme on the hidden variables adopted in a Bohm-inspired quantum network architecture. Thus, the latter becomes in effect a quantum gravity autoencoder for the network materializing emergent gravity.

To develop a network with emergent quantum behavior, we first need to focus on NNs as statistical mechanics entities, and closely examine the statistical physics that may underlie generic machine learning. To start, let us consider the $N \times N$ connectivity matrix **w** and the bias N-vector $\mathbf{w}_0$ as stochastic variables with entries generically denoted q_i, $i=1,\ldots, N+N^2$. The NN state vector **x** is thus updated in discrete time steps according to the usual scheme of *f*-activation:

$$\boldsymbol{x}(t+\tau) = \boldsymbol{f}\left(\boldsymbol{w}\boldsymbol{x}(t) + \boldsymbol{w}_0\right) \tag{5.26}$$

Here the time interval τ represents the overall thermalization time for the state vector $\boldsymbol{x}$, whose entries will be regarded as the hidden variables. To develop the near-equilibrium thermodynamics scheme, let us define a loss function $J(\boldsymbol{x},\boldsymbol{q})$ that penalizes departures from the equilibrium which is achieved as $\boldsymbol{x}(t+\tau) \approx \boldsymbol{x}(t)$ for $t \gg \tau$. If μ represents the "reduced mass" of the network, we get

$$J(\boldsymbol{x},\boldsymbol{q}) = \frac{1}{2} \| \boldsymbol{x} - \boldsymbol{f}\left(\boldsymbol{w}\boldsymbol{x} + \boldsymbol{w}_0\right) \|^2 - \mu \| \boldsymbol{x} \|^2 \tag{5.27}$$

Thus, the statistical thermodynamics near equilibrium stem directly from the canonical partition function

$$Z(\beta,\boldsymbol{q}) = \int \exp\left[-\beta J(\boldsymbol{x},\boldsymbol{q})\right] d^N x \tag{5.28}$$

Following the tenets of statistical mechanics, the partition function yields the Helmholtz free energy for the NN given by:

$$A(\beta,\boldsymbol{q}) = -\beta^{-1} \log\left[Z(\beta,\boldsymbol{q})\right] \tag{5.29}$$

In the specific case of an activation function given by the hyperbolic tangent, the partition function for the NN may be calculated as:

$$Z(\beta,\boldsymbol{q}) = 2\pi^{N/2} \left\{\det\left[\beta \boldsymbol{G}(\boldsymbol{w}) + (1-\beta\mu)\boldsymbol{I}\right]\right\}^{-1/2} \tag{5.30}$$

Where $\boldsymbol{G}(\boldsymbol{w})$ is given by [42]:

$$\boldsymbol{G}(\boldsymbol{w}) = (\boldsymbol{I} - \boldsymbol{f}'\boldsymbol{w})^T (\boldsymbol{I} - \boldsymbol{f}'\boldsymbol{w}) \tag{5.31}$$

where $\boldsymbol{f}'$ is the diagonal matrix of the first derivatives of the activation function with respect to each component of the NN state vector $\boldsymbol{x}$.

When the network is at thermodynamic equilibrium, the average loss $\langle J(\boldsymbol{x},\boldsymbol{q}) \rangle = U(\beta,\boldsymbol{q})$ becomes a minimum, hence the Helmholtz free energy becomes

$$A(\beta,\boldsymbol{q})=\frac{1}{2}\beta^{-1}\left[-N\log(2\pi)+\sum_{\lambda_i>\beta^{-1}}\log(\lambda_i)\right]+\tilde{A}(\beta) \tag{5.32}$$

where λ_i are the eigenvalues of $\mathbf{G}(\boldsymbol{w})$ and $\tilde{A}(\beta)=U(\beta)-\beta^{-1}S(\beta)$ is the thermodynamic Helmholtz free energy of the NN.

To describe the thermodynamic behavior of the NN near equilibrium, we take into account that the entropy production is stationary and introduce the time-dependent probability distribution $p(t,\boldsymbol{q})$ with Shannon entropy $S(t,\boldsymbol{q})=-\int p(t,\boldsymbol{q})\log[p(t,\boldsymbol{q})]d\boldsymbol{q}$. Assuming the learning evolutionary drift follows (i.e., is proportional to) the gradient of the free energy, we get $dq_j/dt=\zeta\partial A/\partial q_j$, hence the minimum entropy production yields the set of constraints

$$\frac{\partial A}{\partial t}+\zeta\left(\frac{\partial A}{\partial q_j}\right)^2=\left\langle\frac{dA(t,\boldsymbol{q})}{dt}\right\rangle_t \tag{5.33}$$

This set of equations in turn begets a minimal action principle defined variationally as

$$\frac{\delta\mathcal{J}(p,A)}{\delta p}=\frac{\delta\mathcal{J}(p,A)}{\delta A}=0, \tag{5.34}$$

where the action $\mathcal{J}(\boldsymbol{q},A)$ becomes

$$\begin{aligned}\mathcal{J}(p,A)=&\int_0^\infty\frac{dS(t,\boldsymbol{q})}{dt}dt\\&+\vartheta\iint_{t=0}^\infty p(t,\boldsymbol{q})\left\{\frac{\partial A}{\partial t}+\sum_{j=1}^{N+N^2}\left[\zeta\left(\frac{\partial A}{\partial q_j}\right)^2\right]-\left\langle\frac{dA(t,\boldsymbol{q})}{dt}\right\rangle_t\right\}dtd\boldsymbol{q}\end{aligned} \tag{5.35}$$

with ϑ denoting the corresponding Lagrange multiplier.

The action may be rewritten taking into account the Fokker-Planck equation satisfied by $p(t,\boldsymbol{q})$:

$$\frac{\partial p}{\partial t}=\frac{\partial}{\partial q_j}\left[\frac{D\partial p}{\partial q_j}-\frac{\zeta\partial A}{\partial q_j}p\right] \tag{5.36}$$

where the parameter D plays the role of diffusion coefficient for the NN dynamics near equilibrium. Taking into account Equation (5.36) in the computation of the time derivative of the Shannon entropy $S(t,\boldsymbol{q})$ enables us to rewrite Equation (5.35), so the action now reads:

$$\mathcal{J}(p,A)=\iint_{t=0}^\infty p(t,\boldsymbol{q})\left\{\vartheta\frac{\partial A}{\partial t}+\sum_{j=1}^{N+N^2}\left[\zeta\frac{\partial^2 A}{\partial q_j^2}-4D\frac{\partial^2 p}{\partial q_j^2}+\vartheta\zeta\left(\frac{\partial A}{\partial q_j}\right)^2\right]-\vartheta\left\langle\frac{dA(t,\boldsymbol{q})}{dt}\right\rangle_t\right\}dtd\boldsymbol{q} \tag{5.37}$$

This action can be written equivalently in the form of a Schrödinger action:

$$\mathcal{J}(p,A)=\iint_{t=0}^{\infty}\Psi^{*}\left[-4D\sum_{j=1}^{N+N^2}\frac{\partial^2}{\partial q_j^{\,2}}-i\eta\frac{\partial}{\partial t}+\Upsilon(\boldsymbol{q})\right]\Psi dtd\boldsymbol{q} \tag{5.38}$$

with $\eta=\left(\frac{4D}{\zeta}\right)^{1/2}$, $¥(\boldsymbol{q})=-\left\langle\frac{dA(t,\boldsymbol{q})}{dt}\right\rangle_t$ and wave function

$$\Psi=p^{1/2}exp\left[i\eta^{-1}A\right] \tag{5.39}$$

Thus, naming $\eta=\hbar$, we obtain the Schrödinger equation describing the state of the system as a particle wave in the q-space $\mathbb{R}^{N\times N}\times\mathbb{R}^{N}$ of trainable variables for the learning process near equilibrium with state $\boldsymbol{x}$-vector entries represented as thermalized hidden variables:

$$i\eta\frac{\partial}{\partial t}\Psi=\left[-4D\nabla^2+\Upsilon(\boldsymbol{q})\right]\Psi \tag{5.40}$$

We have surveyed the statistical thermodynamics of an NN with stochastic trainable variables. The system evolves with the appropriate time coarse-graining associated with the thermalization limit for hidden Bohm variables representing entries in the state vector. Such a system is capable of exhibiting emergent quantum behavior. We are now ready to address the crucial question that underlies the quantum gravity conundrum from the standpoint of AI:

Can this quantum system be regarded as the variational autoencoder of an NN with emergent gravity in a Minkowski space-time?

To address this question, we need to consider the projection $\pi_\tau:\boldsymbol{x}\rightarrow\bar{\boldsymbol{x}}(\boldsymbol{q})$, where $\bar{\boldsymbol{x}}(\boldsymbol{q})$ is the equilibrated state vector of the NN relative to the specific realization $\boldsymbol{q}$ of the stochastic trainable variables that in turn evolve in multiples of the equilibration time τ. In rigorous terms, we get:

$$\bar{\boldsymbol{x}}(\boldsymbol{q})=\int\boldsymbol{x}\exp\left[-\beta J(\boldsymbol{x},\boldsymbol{q})\right]d^N x \tag{5.41}$$

Thus, we are enquiring whether it is possible to treat the non-equilibrium hidden variables (entries) in state vector $\boldsymbol{x}$ by mapping relativistic strings in a NN with an emergent Minkowski space so that the entropy production in such a system is a function of the metric tensor that describes weak interactions between training subsystems of $\boldsymbol{x}$-values. The answer is affirmative because Equations (5.30) and (5.31) may be specialized to the case where the weight vector $\mathbf{w}$ is simply a permutation matrix Ξ with an arbitrary number of cycles [42, 43] so that the matrix $\mathbf{G}$ now becomes

$$\boldsymbol{G}=(\Xi-\boldsymbol{I})^T(\Xi-\boldsymbol{I}). \tag{5.42}$$

Thus, the stochastic NN with partition function

$$Z(\beta,\Xi) = 2\pi^{N/2}\left\{\det\left[\beta(\Xi - \boldsymbol{I})^T(\Xi - \boldsymbol{I}) + (1-\beta\mu)\boldsymbol{I}\right]\right\}^{-1/2} \tag{5.43}$$

represents a quantum gravity autoencoder for the NN with emergent relativistic gravity so that the following diagram becomes commutative:

$$\begin{array}{ccc} \boldsymbol{x}(t) & \overset{\pi_\tau}{\rightarrow} & \overline{\boldsymbol{x}}(t,\Xi) \\ \downarrow F_{ST} & & \downarrow K_{QM} \\ \boldsymbol{x}(t+\tau) & \overset{\pi_\tau}{\rightarrow} & \overline{\boldsymbol{x}}(t+\tau,\Xi) \end{array} \tag{5.44}$$

where F_{ST} and K_{QM} represent respectively the string and quantum flow map.

The emergent quantum behavior was shown to average out or thermalize the hidden variables that have been identified as components of the state $\boldsymbol{x}$-vector of the network. Thus, the trainable variables conforming to the $\boldsymbol{q}$-vector were shown to exhibit a quantum mechanical behavior in an equilibrium regime. We assume without loss of generality that the learning process involves L separate sets of training $\boldsymbol{x}$-vectors with expected values $\overline{\boldsymbol{x}}^l, l = 1,2,\ldots,L$, and these expectation vectors, together with the overall expectation vector (l=0) representing the sum of all L expectation training vectors, are regarded as the hidden variables in the emergent quantum behavior of the trainable $\boldsymbol{q}$-states. On the other hand, the non-equilibrium dynamics of the hidden variables become relevant on timescales much smaller than their thermalization time. These non-equilibrium dynamics are determined by the strength of the weak interactions between vector pairs $\overline{\boldsymbol{x}}^\nu, \overline{\boldsymbol{x}}^\xi, \nu,\xi = 0,1,\ldots,L$ quantified by the tensor $g^i_{\nu\xi}$, where the dummy index i labels each neuron in the system.

To endow the hidden variables with an emergent gravity action, we first describe the non-equilibrium dynamics of the expectation vectors for the training sets. For the sake of transparency, we assume the simplest possible activation function $\boldsymbol{f}$=$\boldsymbol{I}$. Thus, for $\Delta t = \tau_< \ll \tau$, we get to the first order:

$$\overline{x}_i^\mu(t+\tau_<) \approx w_{ij}\overline{x}_j^\mu(t), \mu = 0,1,\ldots,L \tag{5.45}$$

This non-equilibrium scheme yields a tangent bundle according to

$$\frac{\partial \overline{x}_i^\mu}{\partial t} \approx \tau_<^{-1}\left[w_{ij} - \delta_{ij}\right]\overline{x}_j^\mu(t) \tag{5.46}$$

And since $\overline{x}^0 = \sum_{l=1}^{L} \overline{x}^l$, we get

$$g_{\mu\nu}^{i}\frac{\partial\overline{x}_{i}^{\mu}}{\partial t}\frac{\partial\overline{x}_{i}^{\nu}}{\partial t}=0 \tag{5.47}$$

where the metric tensor $g_{\mu\nu}^{i}$ describes the magnitude of the interactions between the hidden variables now cast in terms of the differential geometry of the emergent space-time.

The interactions between different training sets arise from the loss function $J_{<}(\boldsymbol{q})$ that holds for timescales shorter than the equilibration time for the hidden variables. This loss function becomes:

$$J_{<}(\boldsymbol{q})=J_{<}(\boldsymbol{w})=\sum_{l=0}^{L}J_{<}^{(l)}(\boldsymbol{w})=\sum_{l=0}^{L}\sum_{x^{l}\in\mathcal{S}_{l}}\sum_{n=1}^{M=\left[\frac{\tau}{2\tau_{<}}\right]}\|\boldsymbol{x}^{l}\left((n+1)\tau_{<}\right)-\boldsymbol{w}\boldsymbol{x}^{l}\left(n\tau_{<}\right)\|^{2} \tag{5.48}$$

where $\mathcal{S}^{l}\,(l=0,\ldots,L)$ is the l-th training set. Thus, we define the metric tensor as

$$g_{\mu\nu}^{i}=\left|\left[Argmin\left(J_{<}^{(\mu)}\right)-Argmin\left(J_{<}^{(\nu)}\right)\right]_{i}\right|^{-1}\tau^{-1}\sqrt{\int_{t=0}^{\tau}\|\overline{\boldsymbol{x}}^{\mu}(t)-\overline{\boldsymbol{x}}^{\nu}(t)\|^{2}\,dt} \tag{5.49}$$

The gravity action then becomes:

$$\mathcal{J}(\boldsymbol{w})=g_{\mu\nu}^{i}\left[\tau_{<}^{-2}\left\langle x_{i}^{\mu}G_{ij}x_{j}^{\nu}\right\rangle-\frac{\partial\overline{x}_{i}^{\mu}}{\partial t}\frac{\partial\overline{x}_{i}^{\nu}}{\partial t}\right] \tag{5.50}$$

with $\boldsymbol{G}=\boldsymbol{G}(\boldsymbol{w})=(\boldsymbol{I}-\boldsymbol{w})^{T}(\boldsymbol{I}-\boldsymbol{w})$.

Or in Einstein's relativity terms:

$$\mathcal{J}(\boldsymbol{w})=\int d\boldsymbol{X}\sqrt{-g}\,g_{\mu\nu}T^{\mu\nu}, \tag{5.51}$$

where $g=\det(g_{\mu\nu})$ and

$$\sqrt{-g}T^{\mu\nu}=\left[\tau_{<}^{-2}\left\langle\overline{x}_{i}^{\mu}G_{ij}\overline{x}_{j}^{\nu}\right\rangle-\frac{\partial\overline{x}_{i}^{\mu}}{\partial t}\frac{\partial\overline{x}_{i}^{\nu}}{\partial t}\right]\prod_{l=0}^{L}\delta\left(X_{i}^{l}-\overline{x}_{i}^{l}\right) \tag{5.52}$$

Equations (5.48)–(5.52) define the emergent gravity of the neural network arising from the non-equilibrium dynamics of the hidden variables in the quantum mechanical autoencoder.

5.16 RELATIVISTIC STRINGS IN THE QUANTUM PHYSICS OF MACHINE LEARNING

The statistical thermodynamics of machine learning is currently being elucidated by turning nontrainable ($\boldsymbol{x}$) and trainable ($\boldsymbol{q}$) variables into the stochastic variables for the NN and its variational autoencoder, respectively [42]. As demonstrated above, a NN may be endowed with emergent gravity, while its autoencoder is governed by a latent Schrödinger equation, Equation (5.40), thus exhibiting a quantum behavior. To generate this metamodel of quantum gravity, it is necessary to (a) treat the nontrainable variables as hidden variables in the emerging quantum gravity autoencoder, (b) consider a limit where the weight matrix (*w*) becomes a permutation matrix, and (c) treat the hidden variables in a non-equilibrium setting on timescales shorter than thermalization times by generating subsystems of state vectors whose dynamics are described by relativistic strings in an emergent Minkowski space-time. The latent manifold associated with the Minkowski space-time is then obtained by thermalization of the hidden variables. The relativistic strings become enslaved or entrained in the thermalization limit where the nontrainable variables are treated as equilibrated vis-à-vis the trainable variables, and as such they are subsumed in the latent Schrödinger equation via the Helmholtz free energy.

Thus, we may conclude by stating that AI provides a quantum metamodel of gravity, and hence the Big Bang is in all likelihood a quantum event. In this context at least one of the following four statements is valid:

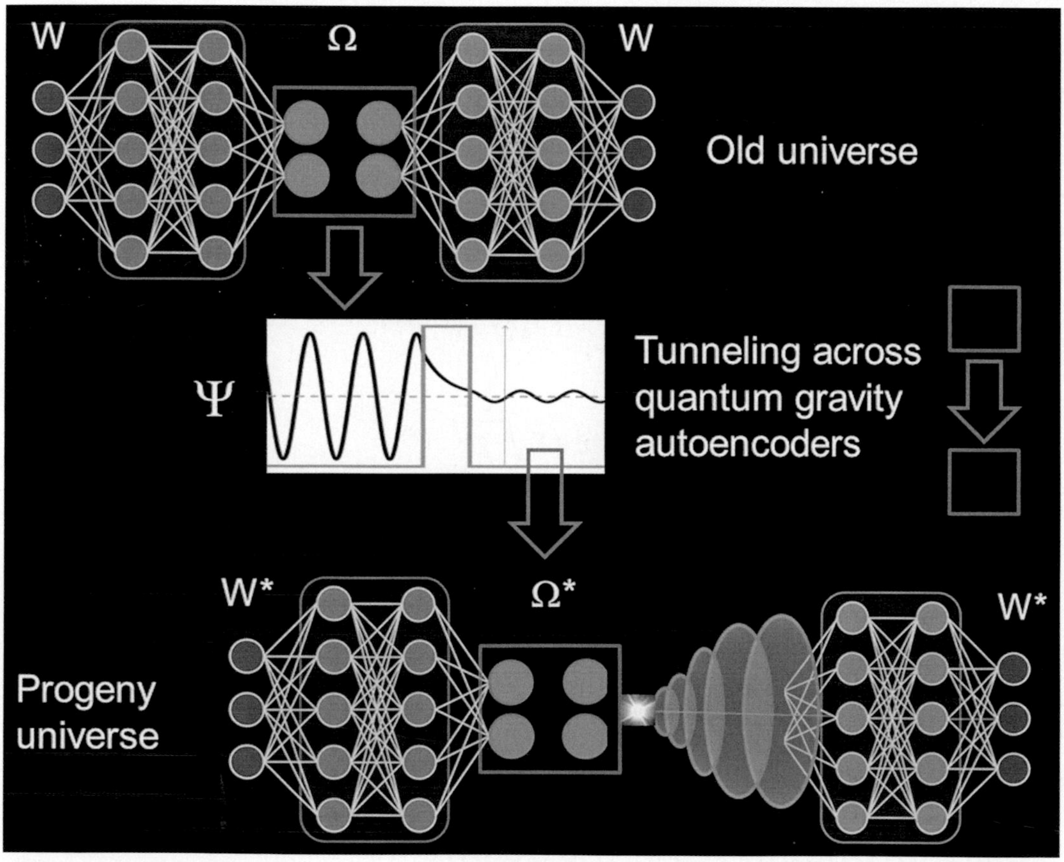

FIGURE 5.18 Schematic representation of a tunneling event across two quantum gravity autoencoders generating the universe (W*, Ω*) as the progeny of an older universe (W, Ω). The amplification of information tunneled to the latent manifold Ω* in a quantum spill-over event materializes as the relativistic decoding of the latent wavefunction, a creation event giving rise to the progeny universe.

A) The laws of quantum mechanics cannot be upheld in the Big Bang setting.

B) A quantum tunneling event generated the universe as a progeny of another universe. The tunneling occurred across the barrier separating two quantum gravity autoencoders with respective latent manifolds Ω and Ω^*. The amplification of the information tunneled to the latent manifold Ω^* was realized as the funneled decoding of the latent wavefunction, giving rise to the progeny universe (Figure 5.18).

C) The *a priori* presence of a primeval observer "Sensus Dei" materialized the Big Bang, which therefore is not a phenomenon-to-be in Wheeler's sense [44] but a realized event.

D) The Big Bang is a phenomenon-to-be in Wheeler's sense and hence we are part of a multiverse, with the universe as a possibility.

5.17 THE UNIVERSE AS A HOLOGRAPHIC AUTOENCODER

This chapter addressed the conundrum of quantum gravity by reductively regarding the universe as the realization of a learning system with stochastic weights and biases where gravity and quantum behavior become emergent properties within the physics of machine learning. To that effect, we explore the possibility of an AI-based construction of a quantum holographic autoencoder, which requires that the emergent quantum behavior arises in a neural network. Once an emergent quantum behavior is shown to become possible within the machine learning system equilibrated on the nontrainable hidden variables, we address the question of developing a relativistic string gravitational scheme on the hidden variables adopted. Thus, the network with equilibrated nontrainable variables becomes in effect a quantum gravity autoencoder for the underlying network exhibiting emergent gravity in the non-equilibrium regime prior to the equilibration of the nontrainable variables. In this way, we build a quantum metamodel for gravity that fulfills at least in part a major imperative for physicists seeking a unified field theory. Furthermore, the physical possibility of tunneling across quantum gravity autoencoders supports the idea that our universe may be the progeny of an older universe [45] that dreamed – or simulated – it.

This is a bold claim yet it may be deconstructed vis-à-vis the main objective of this chapter, which was to describe the behavior of the neural networks in the limit where the bias vector, weight matrix, and state vector of neurons can be modeled as stochastic variables that undergo a learning evolution. As it turns out, this learning evolution, when projected onto the autoencoder, can be described by the time-dependent Schrödinger equation, Equation (5.40), and the time evolution dictated by this equation is compatible with the relativistic decoding enshrined in the commutativity of the diagram presented in Equation (5.44). Taken together, these results have clear implications for the possible emergence of quantum mechanics, general relativity, and mesoscopic observers in neural networks governed by a unified theoretical scheme that adopts two different guises in the two different thermodynamic regimes (cf. Equation (5.44)).

Thus, our construct upholds the controversial view that quantum mechanics may not be a fundamental theory but rather an *ansatz* giving rise to a mathematical tool that allows us to carry out statistical calculations with great efficacy and accuracy in a certain class of dynamical systems. In this guise, emergent quantum mechanics should be derivable from the first principles of statistical mechanics. This is precisely what this chapter has accomplished for a dynamical system consisting of a neural network that is in effect a learning system that contains two different types of degrees of freedom: The trainable bias vector and weight matrix elements, and the nontrainable state vector of neurons, with the latter constituting the hidden variables.

Emergent (or we may say entropic) gravity is also a relatively new area of research, but in this case, the picture is far more nebulous than in emergent quantum mechanics: It is far less clear whether progress has been made, if at all. The main hurdle is that emergent gravity requires also an emergent space, an emergent Lorentz invariance, and an emergent general relativity [46, 47]. To our surprise, the string-theory-based non-equilibrium treatment of the hidden variables in neural networks opened up a window of opportunity to treat the conundrum of emergent gravity in a completely unified fashion that encompasses all three aspects of the problem mentioned above in the context of the learning dynamics. As appears to be the case, a relativistic space-time can indeed emerge from a non-equilibrium evolution of the hidden variables in a manner that is very much akin to string theory [42]. More specifically, as described by Vanchurin [42], if one considers D minimally-interacting subsystems (through bias vector and weight matrix) with average state vectors, then the emergent dynamics can be modeled with relativistic strings in an emergent D + one-dimensional Minkowski space-time. Furthermore, the emergent dynamics may be modeled with the Einstein equations provided the weak subsystem interactions are described by a metric tensor. In this way, a stochastic learning dynamics scheme such as the one proposed in this chapter proved to be instrumental for the equilibration of the emergent space-time that turned out to exhibit a behavior describable by a gravitational theory such as general relativity.

5.18 COSMOLOGICAL TECHNOLOGY USING QUANTUM GRAVITY AUTOENCODERS

The previous discussion addresses one of the biggest problems concerning the history of the universe: What happened before the Big Bang? One may recall that Albert Einstein was never satisfied with the Big Bang scenario itself because he thought that a beginning in time would need to be postulated in a seemingly *ad hoc* manner. This way of doing physics to him was unacceptable, symptomatic of a feeble theory.

After nearly a century since Einstein voiced his skepticism – which extended to quantum mechanics itself – a whole gamut of hypotheses has been formulated regarding our cosmic origin. Perhaps the most sound includes ideas such as the following: (a) The universe sprouts from a quantum vacuum fluctuation, (b) the universe involves infinite cycles of contraction and expansion, (c) the universe was selected through the anthropic principle stemming from the string theory landscape of the multiverse, where every possible event or phenomena is implicitly encompassed, none materialized, and the Big Bang itself is a phenomenon-to-be in Wheeler's sense, and (d) the universe emerged from the collapse of matter in the interior of a black hole that was contained in a progenitor universe.

Be that as it may, none of these ideas can be ascribed full credibility, mainly because none of the theories they stem from has satisfactorily solved the conundrum of quantum gravity. A less explored and more daring possibility put forth in this chapter is that our universe was created in the laboratory by a technologically advanced civilization capable of harnessing the power of quantum gravity autoencoders. This requires a mastery of the physics of learning machines and an ability to craft gravity and quantum behavior as emergent attributes of a stochastic learning system that admits a quantum gravity autoencoder (cf. Equation (5.44)). The underlying theory behind this idea does not portend to solve the quantum gravity conundrum *per se* but at least reconciles the two main forces as emergent in a single learning machine. Furthermore, since our universe is endowed with a flat geometry at zero net energy (Chapter 6), an advanced civilization could have harnessed the power of quantum gravity autoencoders to create a baby universe through quantum tunneling [45] into a second quantum gravity autoencoder acting as a reservoir for the spill-over probability, as schematically depicted in Figure 5.18.

This "emergent matrix" idea of the origin of the universe reconciles the theological need for a "creator," i.e., the primeval quantum observer that would have bestowed reality to the Big Bang by detecting the event, with the secular concept of quantum gravity. As said, while we have not found a cogent theory that conceptually unifies quantum mechanics and gravity, we have a "matrix framework of the universe," a neural network architecture where the two key forces in modern physics are reconciled, so that quantum behavior in the autoencoder can be decoded back as gravity through the holographic map that constitutes the inverse of the canonical projection. Surely a more advanced civilization that masters the quantum gravity autoencoder technology would be able to accomplish the feat of creating baby universes leveraging quantum gravity autoencoders or other equivalent vehicles for quantum tunneling. If so, our universe was not selected for us to dwell in it and bestow reality to the quantum events we are capable of detecting – as upheld by the standard anthropic principle – but rather, it was selected to host civilizations that are much more technologically advanced than we are. These civilizations capable of leveraging the technology required to create progeny universes, be it quantum gravity autoencoders or some unfathomable alternative vehicle efficacious at harnessing quantum tunneling, would be the actual drivers of the cosmic selection process.

By contrast, we are incapable at this time of harnessing technology for cosmological manipulation, and, obviously, we are incapable of recreating the cosmic conditions that led to our existence. In plain words, our civilization is cosmologically still at a rudimentary stage since we do not possess the technology to reproduce the universe that has hosted us for quite a while already. If we were to measure the technological level of a civilization by the ability to recreate or reproduce the astrophysical conditions that led to its existence, we would say that we are at the early stages of development, possibly a low-level civilization, graded class C on a cosmic scale. By contrast, a civilization in the A class rank could recreate the astrophysical conditions that gave rise to its existence, namely produce a baby universe through a quantum-controlled laboratory experiment that leverages a tunneling effect through an appropriately crafted vehicle. A class A civilization would also be able to effectively address related challenges, such as producing a large enough density of dark energy to hold the universe together after its inception, as has already been discussed in the scientific literature [45].

However sound and cogent, the theories on the quantum origins of the universe proposed so far [48] can be subject to a common and basic criticism: *There is no certainty that the universe can be treated as a quantum object.* This chapter heralds an improvement of this state of affairs as the duality enshrined in quantum gravity is realized through a purposely built autoencoder that compresses gravitational string data as emergent quanta.

In accordance with Equation (5.40), if our universe is to become information-compressed within the quantum gravity autoencoder, the following relations must hold: $\zeta = 1/2m$, $D = \hbar^2/8m$, or reciprocally:

$$\hbar = 2\sqrt{\frac{D}{\zeta}}, \tag{5.53}$$

begetting the uncertainty relation adapted for the quantum gravity autoencoder:

$$\Delta E \times \Delta t \sim \sqrt{\frac{D}{\zeta}} \tag{5.54}$$

While the total energy of the NN with emergent gravity is zero (a closed universe has zero energy [49]), the total energy of its quantum gravity autoencoder is $U = -\partial/\partial\beta\,(\log Z(\beta, \boldsymbol{q})) \neq 0$ in accord with the tenets of statistical mechanics. This paradox may be resolved by noting that the observational timescales for the gravitational NN and its autoencoder are different. Thus, in accordance with the uncertainty principle, the total energy of the autoencoder may be zero for a timespan Δt, which is incommensurably shorter than the equilibration time τ for the hidden variables. This implies that the parameter τ needs to be tuned as an architectural determinant of the autoencoder so that the following relation is fulfilled:

$$\tau \gg \Delta t \sim \sqrt{\frac{D}{\zeta}}\left|\frac{\partial}{\partial\beta}\log Z(\beta, \boldsymbol{q})\right|^{-1} \tag{5.55}$$

Thus, the incommensurability of the equilibration timescale relative to the timescales associated with the hidden variables modeled with relativistic strings implies that the universe may be a vacuum quantum fluctuation. This possibility is allowed by the uncertainty principle as described by Equation (5.55).

We should emphasize that the possibility that the universe as intelligible information is actually a vacuum quantum fluctuation is not as far-fetched as it may seem. A simple back-of-the-envelope calculation involving the cosmological constants of our known universe leads to an equivalent result. Thus, the energy $E=mc^2$ of a material object of mass m is actually counterbalanced by its gravitational potential energy $E_g=-GmM/R$, where G is the gravitational constant and M is the net mass of the universe contained within the Hubble ball of radius $R=c/H_0$, where H_0 is the Hubble constant [49]. To prove the previous assertion, we note that the critical minimal mass contained in the Hubble ball of volume $(4/3)\pi(c/H_0)^3$ and required for the universe to be closed is $M=c^3/(2GH_0)$, implying that the gravitational energy E_g compensates the energy E up to a constant of order unity.

The results described in this section pave the way for a cosmological technology that harnesses AI, or more specifically, the power of quantum gravity autoencoders. Thus, two parameters, D and τ, may be tuned to harness the power of AI to manipulate cosmological scales to the point of giving birth to a universe that serves as a metamodel for emergent quantum gravity.

REFERENCES

1. Russell S, Norvig P (2020) *Artificial Intelligence: A Modern Approach*. Pearson, London
2. Kelleher JD (2019) *Deep Learning*. The MIT Press, Cambridge, MA
3. Chollet F (2019) *Deep Learning with Python*. Manning Press, Shelter Island, New York
4. Atienza R (2020) *Advanced Deep Learning with TensorFlow 2 and Keras*, 2nd Edition. Packt Publishing, Birmingham, UK
5. Schmidhuber (2015) Deep learning in neural networks: An overview. *Neural Netw* 61:85–117
6. Fernández A (2016) *Physics at the Biomolecular Interface*. Springer International Publishing, Switzerland
7. Aliper A, Plis S, Artemov A, Ulloa A, Mamoshina P, Zhavoronkov A (2016) Deep learning applications for predicting pharmacological properties of drugs and drug repurposing using transcriptomic data. *Mol Pharm (ACS)* 13: 2524–2530
8. Lavecchia A (2019) Deep learning in drug discovery: Opportunities, challenges and future prospects. *Drug Disc Today* 24: 2017–2032
9. Jiménez J, Škalič M, Martínez-Rosell G, De Fabritiis, G (2018) Kdeep: Protein–ligand absolute binding affinity prediction via 3D-convolutional neural networks. *J Chem Inf Mod* 58: 287–296
10. Stepniewska-Dziubinska MM, Zielenkiewicz P, Siedlecki P (2018) Development and evaluation of a deep learning model for protein–ligand binding affinity prediction. *Bioinformatics* 34: 3666–3674
11. Hassan-Harrirou H, Zhang C, Lemmin T (2020) RosENet: Improving binding affinity prediction by leveraging molecular mechanics energies with an ensemble of 3D convolutional neural networks. *J Chem Inf Model* 60: 2791–2802
12. Brunton SL, Kutz NJ (2019) *Data-Driven Science and Engineering: Machine Learning, Dynamical Systems and Control*. Cambridge University Press, Cambridge, UK
13. Tiumentsev Y, Egorchev M (2019) *Neural Network Modeling and Identification of Dynamical Systems*. Academic Press, San Diego
14. Champion K, Lusch B, Kutz JN, Brunton SL (2019) Data-driven discovery of coordinates and governing equations. *Proc Natl Acad Sci USA* 116: 22445–22451
15. Schmidt M, Lipson H (2009) Distilling free-form natural laws from experimental data. *Science* 324: 81–85
16. Fernández A (1985) Center-manifold extension of the adiabatic-elimination method. *Phys Rev A* 32: 3070–3076

17. Born M, Oppenheimer JR (1927) Zur Quantentheorie der Molekeln. *Ann Physik* 389: 457–484
18. Fernández A (2014) Chemical functionality of interfacial water enveloping nanoscale structural defects in proteins. *J Chem Phys* 140: 221102
19. Fernández A (2021) *Artificial Intelligence Platform for Molecular Targeted Therapy: A Translational Approach*. Chapter 9. World Scientific Publishing, Singapore
20. Fernández A (2020) Deep learning unravels a dynamic hierarchy while empowering molecular dynamics simulations. *Ann Physik* 532: 1900526
21. Brunton SL, Proctor JL, Kutz NJ (2016) Discovering governing equations from data by sparse identification of nonlinear dynamical systems. *Proc Natl Acad Sci USA* 113: 3932–3937
22. Kurzweil R (2006) *The Singularity Is Near: When Humans Transcend Biology*. Penguin, New York
23. de Vries J (2014) *Topological Dynamical Systems: An Introduction to the Dynamics of Continuous Mappings*. De Gruyter, Berlin
24. Duraisamy K, Iaccarino G, Xiao H (2018), Turbulence modeling in the age of data. *Annu Rev Fluid Mech* 51: 357–377
25. Thommen M, Holtkamp W, Rodnina MV (2017) Co-translational protein folding: Progress and methods. *Curr Opin Struct Biol* 42: 83–89
26. Sorokina I, Mushegian A (2018) Modeling protein folding in vivo. *Biology Direct* 13: 13
27. Thurston WP (1997) *Three-dimensional Geometry and Topology*. Princeton University Press, Princeton, NJ
28. Brooks B, Karplus M (1983) Harmonic dynamics of proteins: Normal modes and fluctuations in bovine pancreatic trypsin inhibitor. *Proc Natl Acad Sci USA* 80: 6571–6575
29. Thirumalai D, Lorimer GH, Hyeon C (2020) Iterative annealing mechanism explains the functions of the GroEL and RNA chaperones. *Protein Sci* 29: 360–377
30. Nemytskii VV, Stepanov V (2016) *Qualitative Theory of Differential Equations*. Princeton University Press, Princeton, NJ
31. Arnold VI (1974) *Mathematical Methods of Classical Mechanics*. Springer, Berlin
32. Weisstein EW (2021) Quotient space. From MathWorld: A wolfram web resource. https://mathworld.wolfram.com/QuotientSpace.html
33. Li S, Yang Y (2021) Hierarchical deep learning for data-driven identification of reduced-order models of nonlinear dynamical systems. *Nonlinear Dyn* 105: 3409–3422
34. Maldacena J (1999) The large N limit of superconformal field theories and supergravity. *Int J Theor Phys* 38: 1113
35. Witten E (1998) Anti de Sitter space and holography. *Adv Theor Math Phys* 2: 263–291
36. Bekenstein JD (1973) Black holes and entropy. *Phys Rev D* 7: 2333–2346
37. Fernández A, Sinanoglu O (1982) The lifting of an Inonu-Wigner contraction at the level of universal coverings. *J Math Phys* 23: 2234
38. Hashimoto K, Sugishita S, Tanaka A, Tomiya A (2018) Deep learning and the AdS/CFT correspondence. *Phys Rev D* 98: 046019
39. Matsueda H, Ishihara M, Hashizume Y (2013) Tensor network and a black hole. *Phys Rev D* 87: 066002
40. Weinberg S (2008) *Cosmology*. Oxford University Press, Oxford
41. Bohm D (1962) A suggested interpretation of the quantum theory in terms of hidden variables I. *Phys Rev* 86: 166–179
42. Vanchurin V (2021) Toward a theory of machine learning. *Mach Learn Sci Technol* 2: 036012
43. Gubser SS (2010) *The Little Book of String Theory*. Princeton University Press, Princeton, NJ
44. Wheeler JA, Zurek WH (2014) *Quantum Theory and Measurement*. Princeton University Press, Princeton, NJ
45. Farhi E, Guth A, Guven J (1990) Is it possible to create a universe in the laboratory by quantum tunneling? *Nuc Phys B* 339: 417–490

46. Vanchurin V (2018) Covariant information theory and emergent gravity. *Int J Mod Phys A* 33: 1845019
47. Bednik G, Pujolas O, Sibiryakov S (2013) Emergent Lorentz invariance from strong dynamics: Holographic examples. *J High Energy Phys* 11: 64
48. Loeb A (2021) Was our universe created in a laboratory? *Scientific American*. October 15, 2021. https://www.scientificamerican.com/article/was-our-universe-created-in-a-laboratory/
49. Harrison ER (2003) *Masks of the Universe*. Cambridge University Press, Cambridge, UK

CHAPTER 6

Querying Artificial Intelligence on Dark Matter and Dark Energy

Quintessential Reverse Engineering of the Standard Model

No phenomenon is a physical phenomenon
until it is an observed phenomenon.

– JOHN A. WHEELER

SUMMARY

As we examine the cosmos at very large scales, basic tenets of physics crumble under the weight of contradicting evidence stemming from the postulated existence of dark matter (DM) and dark energy (DE). This chapter addresses the problem and is meant to help mitigate the crisis. It resorts to artificial intelligence (AI) for answers and describes the outcome of this quest in terms of an ur-universe that incorporates an extra dimension to encode Einstein's space-time as a latent manifold.

DM is now believed to have arisen during the creation of elementary particles in an early universe. In contrast with our present-day "flat" universe, this early universe was endowed with extreme geometric curvature and, consequently, with special quantization rules. To begin to validate this picture, an AI platform is leveraged that has been previously designed as an autoencoder for dynamical systems. This platform is capable

DOI: 10.1201/9781003385950-6

of reverse engineering the action principles that underlie the Standard Model of elementary particle physics. The deconstruction treats Einstein's 4D space-time as a "latent space" that gets decoded onto a higher dimensional space. The latter is endowed with an extra rolled-up dimension that spans the quark cross-section, which is the smallest material dimension estimated to lie within the attometer (10^{-18}m) scale. It turns out that this compact fifth dimension stores stationary wave matter with a rest mass that matches the vacuum expectation value of the Higgs boson. This result enabled us to estimate elementary particle masses with significant precision by an AI-based decoding of the elementary particle fields along the fifth dimension. The results point to the existence of an ur-Higgs boson in the early universe, specifically at the beginning of the "electroweak epoch," whose kinetic energy is not geometrically diluted along the standard 4D dimensions. This ur-Higgs and its heavier quantum partners are identified by AI as DM. The results enable us to characterize DM vis-à-vis the geometric dilution of gravity shown to have taken place as the universe flattened and expanded to present-day levels while remaining a compact borderless manifold.

The results obtained by AI and reported in this chapter may be summarized as follows:

- The universe topology is "revealed" by the CMB fluctuation spectrum. Unlike the Euclidean topology, this topology is compatible with the Big Bang scenario for universe evolution.
- AI can trace the origin of dark matter to the universe's evolution by reverse-engineering the Standard Model to incorporate a circular dormant dimension.
- On the dormant compact dimension with quark-size radius q, a de Broglie matter-wave has energy equal to the vacuum expectation value of the Higgs field. This is the ur-Higgs elementary particle, a good candidate for dark matter for the following reasons:
  - The incorporation of a fifth dimension is supported by evidence: SM elementary particle masses can be predicted through geometric dilution of the ur-Higgs.
  - The ur-Higgs formed in an early universe ($\sim 10^{-27}$s after the Big Bang).
  - It is massive and cold ($v << c$).
  - It does not decay through communication with SM gauge bosons.
  - It is only interactive with the SM via gravity.
  - It endows other ur-particles with mass through ur-boson-conveyed geometric dilution.

On the other hand, DE is generated by quantum vacuum fluctuations exciting the quintessential field of the ur-Higgs and becoming stored as DM whenever the wavelength of the excitation fits the stationary wave condition along the dormant dimension.

6.1 A QUINTESSENTIAL AUTOENCODER TO DECODE THE STANDARD MODEL

As shown in the scheme in Eq. 4.15 (Chapter 4) and in Chapter 5, the concept of diagram commutativity is central to the encoding of a dynamical system into its latent simplified dynamics and, reciprocally, to the decoding of the latent dynamics onto a space that incorporates dormant dimensions [1]. This scheme is now adapted to decode the SM onto a space with an extra dimension. Thus, the task of decoding the SM defined on the four-dimensional space-time W/~, onto a five-dimensional compact multiply connected manifold W requires a "quintessential" (i.e., "fifth essence") autoencoder operating in reverse. Typically, autoencoders simplify the dynamics to retain the latent coordinates [1]. In the case of interest to elementary particle physics, however, tangible observable processes are assumed to be taking place in a quotient space of equivalent classes modulo a dormant dimension, and the goal of the quintessential autoencoder is to decode the SM onto the space that incorporates the dormant fifth dimension (the "fifth essence").

This reverse quintessential autoencoder lifts the flow $\mathcal{F}: W/\sim \to W/\sim$ onto a flow $\tilde{\mathcal{F}}: W \to W$. The lifting is compatible if and only if it satisfies the commutativity relation: $\mathcal{F} \circ \pi = \pi \circ \tilde{\mathcal{F}}$, with $\pi: W \to W/\sim$ denoting the canonical projection that assigns each point in W to its equivalence class in $W/\sim$. Since we are introducing autoencoders for dynamical systems [1], it becomes imperative to cast the SM as a dynamical system. This prompts the question: How do we cast the flow $\mathcal{F}: W/\sim \to W/\sim$ to represent a process described by the SM?

To answer the question, we may regard the interaction processes determined by the SM as transformations created by a time-dependent operator. For each elementary process, this operator has a time step τ associated with it, and this time step is precisely the lifetime of the gauge boson that communicates the force acting on the elementary particle and causes the elementary particle transformation [2]. Thus, the time step is obtained from the uncertainty principle: $\sim \hbar/2M_B c^2$, where M_B is the rest mass of the relevant gauge boson. Thus, the autoencoder described in Chapter 5 will be operated in reverse in order to decode the SM by incorporating the dormant dimension shown to store gravity in a reverse engineering scheme that fits the Big Bang scenario.

The NN for the quintessential autoencoder (AE) is constructed in such a way that each input node represents the elementary particle field value (for simplicity assumed to be a complex number) in an equivalence class contained in $W/\sim$. The grid of nodes or equivalence classes has a mesh determining the level of resolution, and each equivalence class is taken modulo the dormant coordinate, implying that it is represented by the four latent coordinates of the standard four-dimensional space-time. Since the autoencoder is set up to be working in reverse, the output nodes represent the values of the decoded elementary particle field on points in the quintessential space W. Thus we have the following

Lemma 6.1. The elementary particle field ϕ is decoded by the quintessential autoencoder as the ur-particle field $\tilde{\phi}$ if and only if $\exists \gamma_\phi$:

$$\phi \circ \pi = \gamma_\phi \circ \tilde{\phi}; \quad \gamma_\phi : \mathbb{C} \to \mathbb{C} \tag{6.1}$$

The proof follows from the commutativity of the following diagram that would make the decoding compatible with the field defined on the four-dimensional space-time:

$$W \to W/\sim$$

$$\downarrow \tilde{\phi} \quad \downarrow \phi$$

$$\mathbb{C} \to \mathbb{C} \tag{6.2}$$

The field for an elementary particle with Lagrangian $\mathcal{L}$ satisfies the Euler-Lagrange equation:

$$\partial_\phi \mathcal{L} = \partial_\mu \frac{\partial \mathcal{L}}{\partial(\partial_\mu \phi)} \tag{6.3}$$

while the decoded field of the ur-particle obeys the relation

$$\partial_{\tilde{\phi}} \tilde{\mathcal{L}} = \partial_\nu \frac{\partial \tilde{\mathcal{L}}}{\partial(\partial_\nu \tilde{\phi})}, \tag{6.4}$$

with its respective decoded Lagrangian $\tilde{\mathcal{L}}$ fulfilling the relation

$$\int \tilde{\mathcal{L}}(\tilde{\phi}) d\mathbf{x} dy_5 = \int \mathcal{L}(\phi) d\mathbf{x} + \left[\int \left| \int (\tilde{\phi} - \phi) d\mathbf{x} \right|^2 dy_5 \right]^{1/2} \tag{6.5}$$

Heretofore, the dormant coordinate is denoted y_5.

Equation (6.5) is central to the process of quintessential decoding of the SM. It indicates that the field $\tilde{\phi}$ of the ur-particle is always the minimal smooth extension of the elementary particle field ϕ over the fifth coordinate, while the compactness of the latter coordinate ensures the convergence of the right term in the right-hand side of Equation (6.5).

In principle, we do not and cannot *a priori* specify the radius r_0 of the circular dormant coordinate because this parameter, together with the geometric dilution parameter (or α pitch), must be determined in the optimization of the pair $(\tilde{\phi}, \gamma_\phi)$ in accord with the loss function $J(\alpha, r_0)$ associated to the diagram commutativity (Figure 6.1):

$$J(\alpha, r_0) = \sum_{y \in M_\alpha} \frac{1}{|M_\alpha|} \| (\phi \circ \pi) y - (\gamma_\phi \circ \tilde{\phi}) y \|^2 \tag{6.6}$$

$$(\alpha, r_0)(\phi) = Arg\, Min\, J \tag{6.7}$$

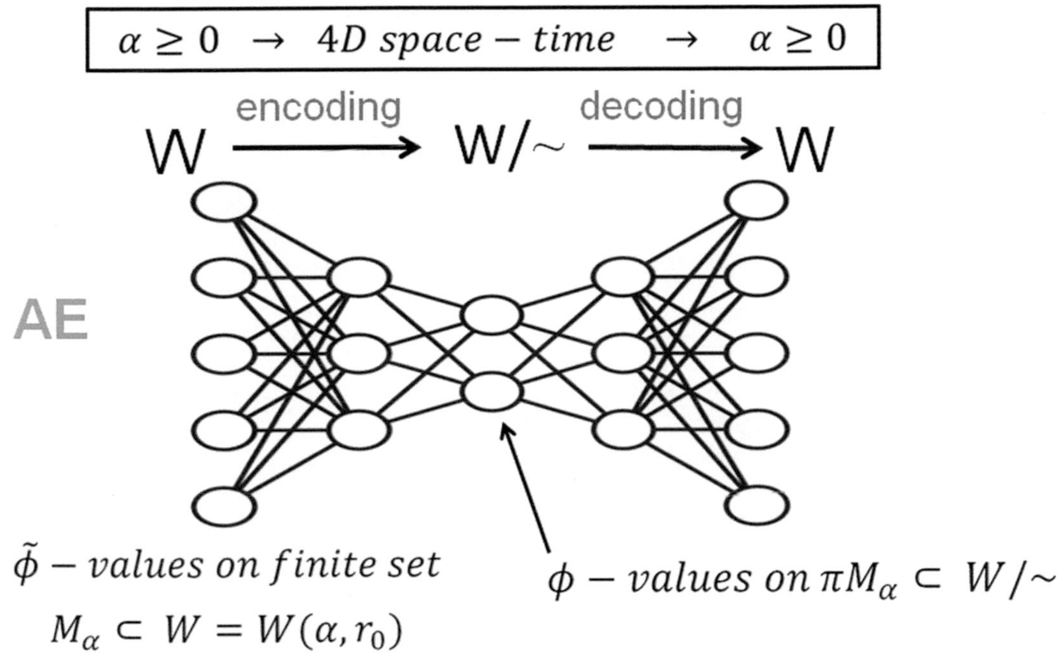

$$J(\alpha, r_0) = \sum_{y \in M_\alpha} \frac{1}{|M_\alpha|} \left\| (\phi \circ \pi) y - (\gamma_\phi \circ \tilde{\phi}) y \right\|^2$$

$$(\alpha, r_0)(\phi) = Arg\ Min\ J$$

FIGURE 6.1 The quintessential autoencoder (AE) decoding an SM particle field ϕ on the four-dimensional space-time that is regarded as the latent manifold *W/~* of the quintessential space *W*. A loss function is parametrically determined by the pitch α and dormant dimension r_0, whose optimization yields the decoded field $\tilde{\phi}$.

In practice, we shall see that the optimization dictated by the loss function attributes a unique geometric dilution value to each elementary particle in the SM, while r_0 remains constant and equal to the quark value q in the attometer range [3]. Assuming and anticipating $r_0=q$, the corresponding wave functions for ur-particles become parametrically dependent on geometric dilution.

By definition, the ur-Higgs boson mass is stored completely in the dormant dimension. Therefore, the wavefunction Υ for the ur-Higgs boson is associated with $\alpha=0$ (no geometric dilution) and its expectation energy becomes $\left\langle \Upsilon \middle| \widetilde{\mathcal{H}_H} \middle| \Upsilon \right\rangle = 246G\ eV$, where $\widetilde{\mathcal{H}_H}$ denotes the decoded Hamiltonian for the ur-Higgs boson. On the other hand, the wave function Ψ_H for the decoded Higgs boson satisfies $\langle \Psi_H | \Upsilon \rangle = \cos\alpha^*$, where $\alpha^* = 59°27' \approx \pi/3$. In general, for an ur-particle ζ with associated pitch α_ζ, we obtain $\langle \Psi_\zeta | \Upsilon \rangle = \cos\alpha_\zeta$. These observations are depicted in Figure 6.2.

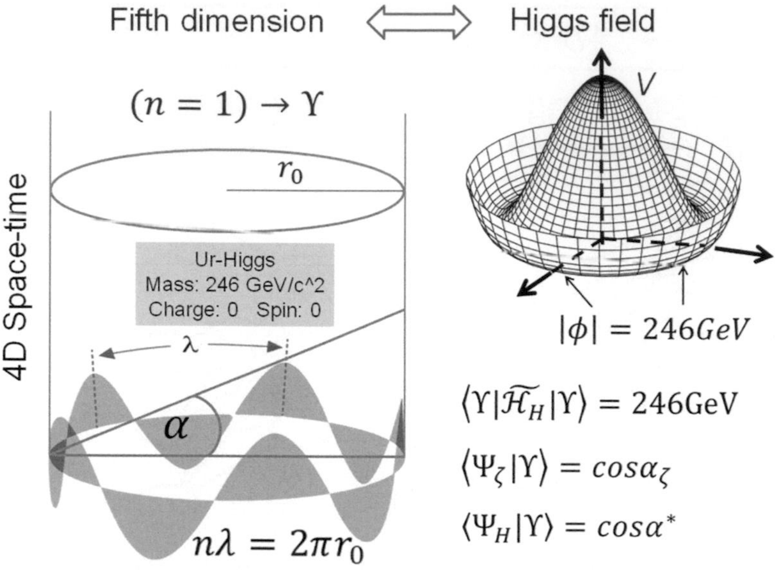

FIGURE 6.2 The dormant fifth dimension in the quintessential decoding of space-time. The fifth dimension represents the smallest material scale known, which corresponds to the quark cross-section, established to be in the attometer range [3]. The lowest energy stationary (de Broglie) matter-wave fully stored in the dormant dimension and named ur-Higgs has a rest mass energy corresponding to the vacuum expectation value of the Higgs boson field $|\phi| \approx 246GeV$.

6.2 AI-ENABLED REVERSE ENGINEERING OF THE STANDARD MODEL THROUGH QUINTESSENTIAL DECODING OF AN EARLY UNIVERSE

In order to compute the quintessential decoding of an elementary particle field to generate the ur-particle field, the reverse autoencoder changes the metric of each Euclidean coordinate in *W*, compressing the metric asymptotically as we approach infinity. This is a purely AI move, completely unsupervised and uninstructed, and compactifies the space as it rolls up each Euclidean dimension onto a circle (Figure 6.3). The Euclidean metric on the line is changed into a new metric obtained by mapping a circle osculating the line so that the distance between two points along the line is now evaluated as the length of the arc sustained between the projected points on the circle, as shown in Figure 6.3.

This new metric essentially maps W homeomorphically onto a five-dimensional torus, so that states representing ur-particles become quantized in W with five quantum numbers due to the requirement to generate stationary de Broglie matter waves in the five-dimensional torus. Thus, quantization rules apply so that the wavelength λ for an elementary particle with pitch α must now fulfill the stationary-wave equation $\lambda \cos\alpha = \frac{2\pi r_0}{n}$ for some quantum number (integer) *n* and simultaneously fulfill four stationary-wave equations of the form $\lambda \sin\alpha = \frac{2\pi r}{n'}$ for quantum numbers *n'* for each coordinate in the four-dimensional space-time endowed with the metric inherited from projection onto a circle of

AI pursues an "early universe" decoding

Changes the metric of W:
d(a,b) = length arc (a',b')
~ compression at infinity
~ euclidean metric
in the limit → 0

"Early universe" quantization
(stability in the spatial 4-torus)

$$\lambda \cos \alpha = \frac{2\pi r_0}{n}$$

$$\lambda \sin \alpha = \frac{2\pi r}{n'}$$

$2\pi r_0 = 5.03656 \times 10^{-18} m$

$$tg\alpha = \left(\frac{n}{n'}\right)\left(\frac{r}{r_0}\right)$$

$$W \xrightarrow{\pi} W/\sim$$

$$\tilde{\phi} \downarrow \qquad \downarrow \phi$$

$$\mathbb{C} \longrightarrow \mathbb{C} \to \mathbb{C}$$
$$\gamma\phi$$

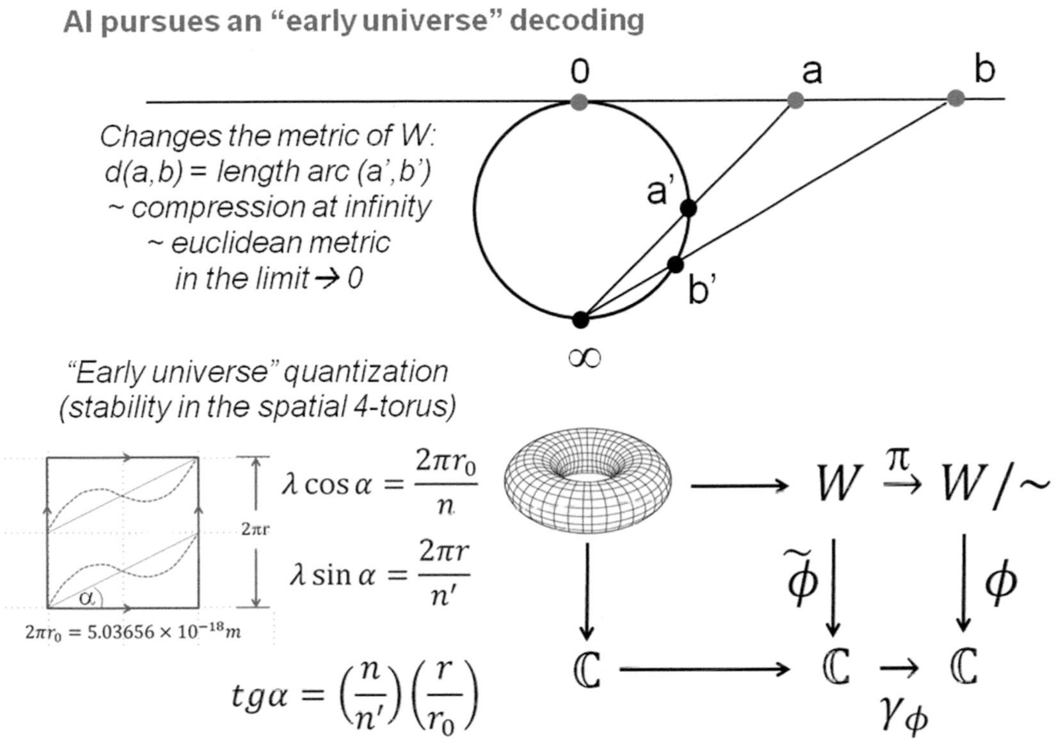

FIGURE 6.3 Early-universe toroidal decoding of the four-dimensional space-time regarded as the latent space (W/~) for the quintessential space W homeomorphic to an "early-universe" five-dimensional torus. The quintessential toroidal decoding of the particle field introduces a quantization required to avoid destructive interference on the five-dimensional torus. The quantization of the ur-field is thus defined by four independent quantum numbers that are encoded into the standard particle attributes within the Standard Model (charge, mass, and spin) plus an extra attribute known as geometric dilution, only apparent in the quintessential decoded space W.

radius r (Figure 6.3). This implies that geometric dilution α, quantum number ratio $\left(\frac{n}{n'}\right)$, and aspect ratio $\left(\frac{r}{r_0}\right)$ are interrelated according to the four fundamental relations:

$$tg\alpha = \left(\frac{n}{n'}\right)\left(\frac{r}{r_0}\right) \tag{6.8}$$

For example, for the Higgs boson at $(n,n')=(1,1)$ with $\alpha = 59°27'$ ($tg\alpha = 1.69428$), we get the aspect ratio $\left(\frac{r}{r_0}\right) = 1.69428$. Since aspects ratios are invariants of the five-dimensional torus and the pitch is fixed for each ur-particle, the set of five quantum numbers $(n, n'_1, n'_2, n'_3, n'_4)$ determines the quintessential quantized decoding of each elementary particle in the SM. The decoding of each elementary particle field through the toroidal quantization of its ur-field in the quintessential space W is thus represented in Figure 6.3. The four elementary

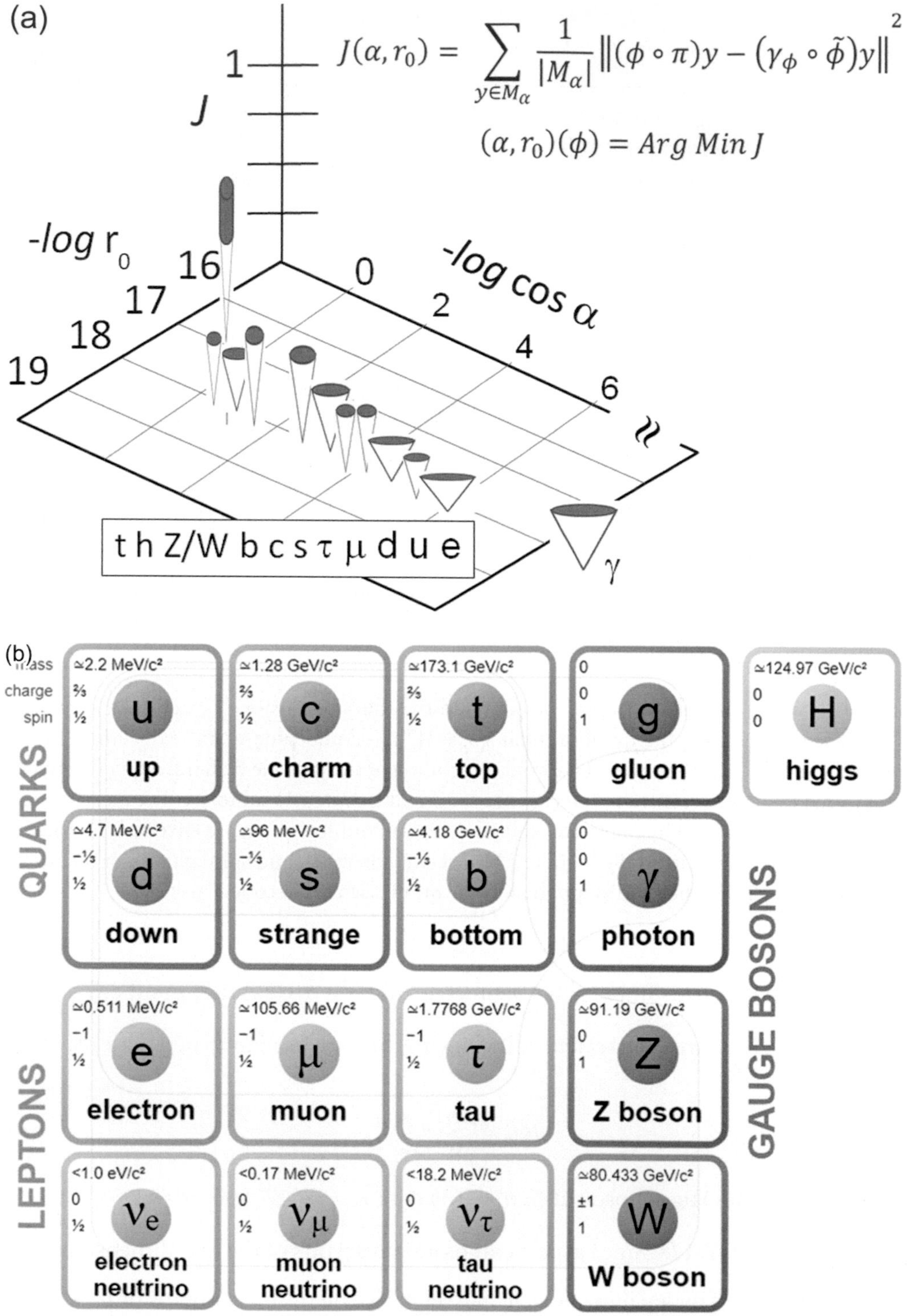

FIGURE 6.4 Pitch and dormant dimension (α, r_0) for the optimization of the loss function $J(\alpha, r_0)$ for the quintessential autoencoder that decodes each elementary particle field in the Standard Model. (a) Loss function optimization in (J, α, r_0)-space for each elementary particle denoted following notation in (b). (b) Table of elementary particles in the Standard Model.

particle attributes in the SM, mass, geometric dilution, charge, and spin represent a specific encoding of the four independent quantum numbers in the toroidal quantization described by the four relations in Equation (6.8).

The geometric dilution associated with each elementary particle in the SM is thus obtained by computing the ur-field as the optimum quintessential decoding of the elementary particle field that minimizes the loss function in fulfillment of Equations (6.6) and (6.7). This decoded field minimizes the quintessential Lagrangian defined by Equation (6.5). The results are presented in Figure 6.4.

The decoding of the SM is fully implemented by ensuring through parametric optimization the commutativity of the dual action diagram presented in Figure 6.5. To introduce this higher level of description some notation becomes necessary: The space of smooth scalar fields for ur-particles is $C^2(W,\mathbb{C})$, while the space $C^2(W/\sim,\mathbb{C})$ contains the elementary particle scalar fields in the SM. The respective dual spaces $C^2(W,\mathbb{C})^*$ and $C^2(W/\sim,\mathbb{C})^*$

$$\boxed{\alpha \geq 0 \qquad \rightarrow \qquad 4D\ space - time \qquad \rightarrow \qquad \alpha \geq 0}$$

$$\widetilde{\mathfrak{A}} \in C^2(W,\mathbb{C})^* \xrightarrow{\ \pi^*\ } \mathfrak{A} \in C^2(W/\sim,\mathbb{C})^* \xrightarrow{\ (\pi^{-1})^*\ } C^2(W,\mathbb{C})^* \in \widetilde{\mathfrak{A}}$$

$$\Gamma_{\widetilde{\mathcal{L}_0}} \Bigg\downarrow \qquad \boxed{\begin{array}{l}\Gamma_{\mathcal{L}_0} \circ \pi^* = \\ = \pi^* \circ \Gamma_{\widetilde{\mathcal{L}_0}}\end{array}} \ \Gamma_{\mathcal{L}_0}\Bigg\downarrow \qquad \boxed{\begin{array}{l}\Gamma_{\widetilde{\mathcal{L}_0}} \circ (\pi^{-1})^* = \\ = (\pi^{-1})^* \circ \Gamma_{\mathcal{L}_0}\end{array}} \ \Gamma_{\widetilde{\mathcal{L}_0}}\Bigg\downarrow$$

$$\widetilde{\mathfrak{A}'} \in C^2(W,\mathbb{C})^* \xrightarrow{\ \pi^*\ } \mathfrak{A}' \in C^2(W/\sim,\mathbb{C})^* \xrightarrow{\ (\pi^{-1})^*\ } C^2(W,\mathbb{C})^* \in \widetilde{\mathfrak{A}'}$$

$$\mathfrak{A} = \mathfrak{A}(\mathcal{L}) = \int \mathcal{L}(\phi,\partial_\mu\phi)dx^\mu \qquad \widetilde{\mathfrak{A}} = \int \widetilde{\mathcal{L}}(\widetilde{\phi},\partial_\nu\widetilde{\phi})dy^\nu$$

$$\Gamma_{\mathcal{L}_0}(\mathfrak{A}) = \mathfrak{A}' = \mathfrak{A}(\mathcal{L}')\ where$$

$$\int [\mathcal{L} + \mathcal{L}_0]dx = \int [\mathcal{L}' + \mathcal{L}'_0]dx'$$

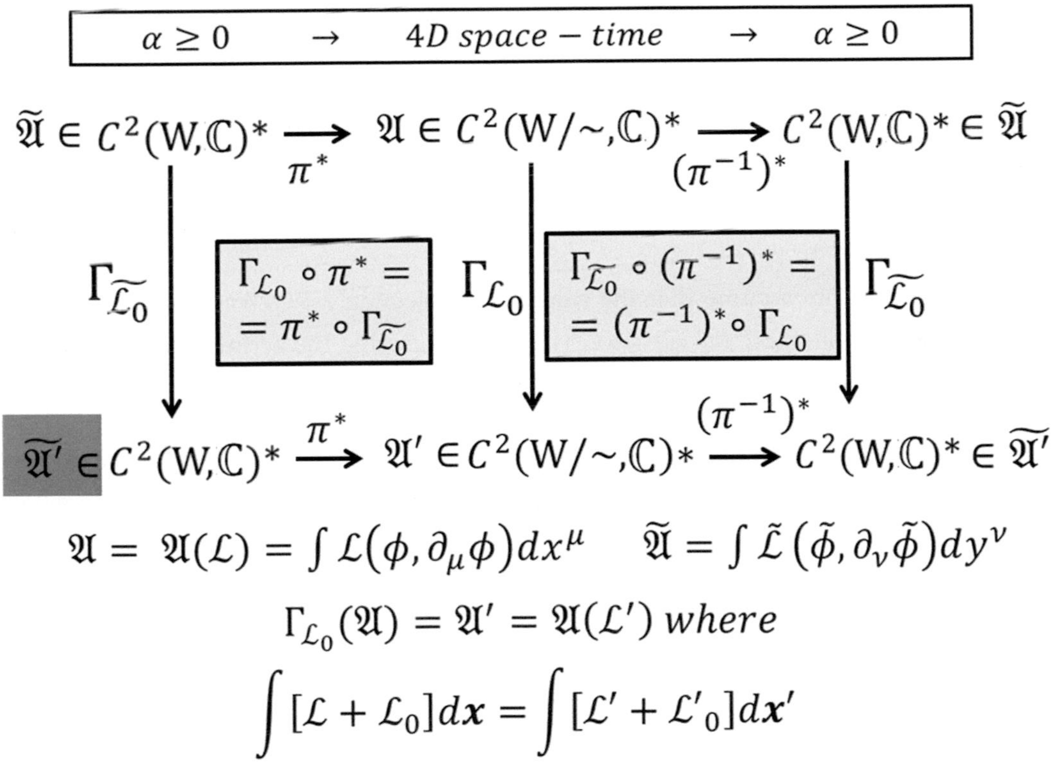

FIGURE 6.5 Quintessential decoding of the Standard Model implemented by ensuring the commutativity of the action diagram through parametric optimization. In this decoding context dual to that represented in Figure 6.1, the gauge boson with Lagrangian $\mathcal{L}_0$ is represented as a map $\Gamma_{\mathcal{L}_0} : C^2(W/\sim,\mathbb{C})^* \rightarrow C^2(W/\sim,\mathbb{C})^*$ defined over action space as $\Gamma_{\mathcal{L}_0}\mathfrak{A}(\mathcal{L}) = \mathfrak{A}(\mathcal{L}')$. Similarly, the corresponding decoded ur-boson is represented by the map $\Gamma_{\widetilde{\mathcal{L}}} : C^2(W,\mathbb{C})^* \rightarrow C^2(W,\mathbb{C})^*$ defined as $\Gamma_{\widetilde{\mathcal{L}}}\widetilde{\mathfrak{A}}(\widetilde{\mathcal{L}}) = \widetilde{\mathfrak{A}}(\widetilde{\mathcal{L}}')$. So the consistent decoding of the SM requires that the following commutativity relation is valid for any generic boson with Lagrangian $\mathcal{L}_0 : \Gamma_{\mathcal{L}_0} \circ \pi^* = \pi^* \circ \Gamma_{\widetilde{\mathcal{L}_0}}$, where π^* is the dual map of the canonical projection of the quintessential space W onto the quotient (latent) space $W/\sim$.

contain the actions defined by functionals-actions of the type: $\tilde{\mathfrak{A}}(\tilde{\mathcal{L}})=\int\tilde{\mathcal{L}}(\tilde{\phi},\partial_\nu\tilde{\phi})\,d\boldsymbol{y}^\nu$ and $\mathfrak{A}(\mathcal{L})=\int\mathcal{L}(\phi,\partial_\mu\phi)\,d\boldsymbol{x}^\mu$, respectively. As described in Chapter 4, an elementary particle Lagrangian $\mathcal{L}$ becomes transformed into another elementary particle Lagrangian $\mathcal{L}$' when the first elementary particle interacts with another elementary particle (gauge boson) with Lagrangian $\mathcal{L}_0$. These interactions define the flow in action space that enables the SM to be represented as a dynamical system with the propagator time step defined as the lifetime of the gauge boson with associated Lagrangian $\mathcal{L}_0$ that communicates the force acting on the elementary particle and induces its transformation. Thus, the SM may be decoded at the action level using dual autoencoders adapted to dynamical systems of the type described in Chapter 5 but acting in reverse, where the four-dimensional space-time represents the latent space.

In the dual context of this decoding, the gauge boson with Lagrangian $\mathcal{L}_0$ is represented as a map $\Gamma_{\mathcal{L}_0}:C^2(W/\sim,\mathbb{C})^*\rightarrow C^2(W/\sim,\mathbb{C})^*$ defined as $\Gamma_{\mathcal{L}_0}\mathfrak{A}(\mathcal{L})=\mathfrak{A}(\mathcal{L}')$. Similarly, the corresponding ur-boson is represented in this dual space by the map $\Gamma_{\widetilde{\mathcal{L}_0}}:C^2(W,\mathbb{C})^*\rightarrow C^2(W,\mathbb{C})^*$ defined as $\Gamma_{\widetilde{\mathcal{L}_0}}\tilde{\mathfrak{A}}(\tilde{\mathcal{L}})=\tilde{\mathfrak{A}}(\widetilde{\mathcal{L}'})$. So the decoding of the SM becomes operational when it is also validated at the level of actions, implying that the following commutativity relation is valid for any gauge boson with generic Lagrangian denoted $\mathcal{L}_0$:

$$\Gamma_{\mathcal{L}_0}\circ\pi^*=\pi^*\circ\Gamma_{\widetilde{\mathcal{L}_0}} \tag{6.9}$$

In Equation (6.9), we assume that the canonical projection $\pi:W\rightarrow W/\sim$ induces the dual canonical projection $\pi^*:C^2(W,\mathbb{C})^*\rightarrow C^2(W/\sim,\mathbb{C})^*$ mapping the action space for W onto the action space for the SM.

Geometric dilution parametrized by the pitch angle α may be communicated gravitationally in the quintessential space W through the respective ur-boson α, as schematically represented in the Feynmann diagram [2] displayed in Figure 6.6. Thus, the ur-Higgs may decay into a geometrically diluted ur-particle with energy $E_\alpha=(c\hbar/r_0)\cos\alpha$ and thereby transfer the geometrically diluted mass m (expressed in GeV/c^2) satisfying the equation $\cos\alpha=(246-m)/246$. The mass is captured by an ur-particle with rest mass m' that gets transformed into an ur-particle with mass $m+m'$ upon interaction with the ur-boson α that has been emitted by or has interacted with the ur-Higgs boson.

6.3 VALIDATING THE QUINTESSENTIAL DECODING OF ELEMENTARY PARTICLES AS UR-PARTICLES

The reverse quintessential autoencoder decodes the elementary particle fields in the SM as ur-particles with masses determined by the geometric dilution of the ur-Higgs stationary wave stored on the dormant fifth coordinate. The geometric dilution may be directly determined from the pitch of the wavefront associated with the quintessentially decoded wave function for the ur-particle. Wave functions for elementary particles are decoded as maps $\Psi:\mathrm{W}\rightarrow\mathbb{C}$ following the same tenets that apply to the decoding of particle fields (Section 6.2). The procedure is shown in Section A.2 of the Appendix. The decoding of the

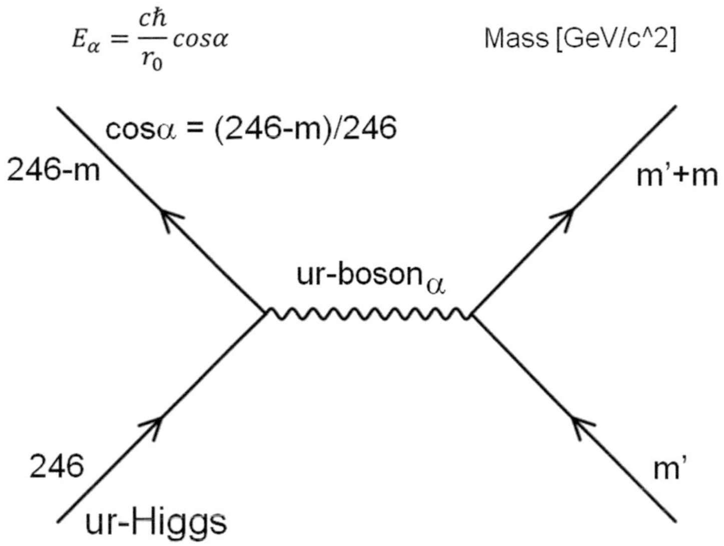

FIGURE 6.6 Communication of geometric dilution in the quintessential space *W* represented as a Feynman diagram (cf. [2]), involving an ur-particle with mass *m'*, the ur-Higgs boson, and an ur-gauge boson associated with the pitch value α.

photon (γ) has infinite dilution (υ=–logcosα=∞, α=π/2), yielding a zero mass (Figure 6.4), and hence this gauge boson communicating the electromagnetic force travels at the speed of light. Thus, the wavefunction $\Psi_\gamma : \mathrm{W} \to \mathbb{C}$ for the ur-photon is orthogonal to that of the ur-Higgs $\langle \Psi_\gamma \mid \Upsilon \rangle = \cos(\pi/2) = 0$. At the opposite end of the spectrum, the heaviest elementary particle, the top quark (t) gets decoded into the ur-t which has the lowest geometric dilution of all decoded elementary particles in the SM: $\upsilon = -\log\cos\alpha = 0.153, \alpha \approx \pi/4$. This yields $\langle \Psi_\mathrm{t} \mid \Upsilon \rangle = 0.703$ a good approximation to the ratio between elementary particle mass and Higgs vacuum expectation value (*v*): $\frac{m}{v} = \frac{173.1}{246} \approx \langle \Psi_\mathrm{t} \mid \Upsilon. \rangle$ The following relation is valid for all gauge bosons in the SM:

$$\frac{m_\zeta}{v} \approx \langle \Psi_\zeta \mid \Upsilon \rangle \tag{6.10}$$

In the case of fermions, the relation becomes

$$g_\zeta \approx \langle \Psi_\zeta \mid \Upsilon \rangle, \tag{6.11}$$

where g_ζ is the respective Yukawa coupling constant [2]. The Yukawa coupling to the Higgs field can thus be interpreted as an effective geometric dilution parameter for ur-fermions in quintessential space as it follows from the relation *g*=cosα.

A linear plotting $\frac{m_\zeta}{v}, g_\zeta - \langle \Psi_\zeta \mid \Upsilon \rangle$ yields R^2=0.88 as shown in Figure 6.7 for all elementary particles in the SM. The resulting prediction of masses is based on the geometric dilution that optimizes the loss function *J*(α,*q*) for the decoding of each particle field into its

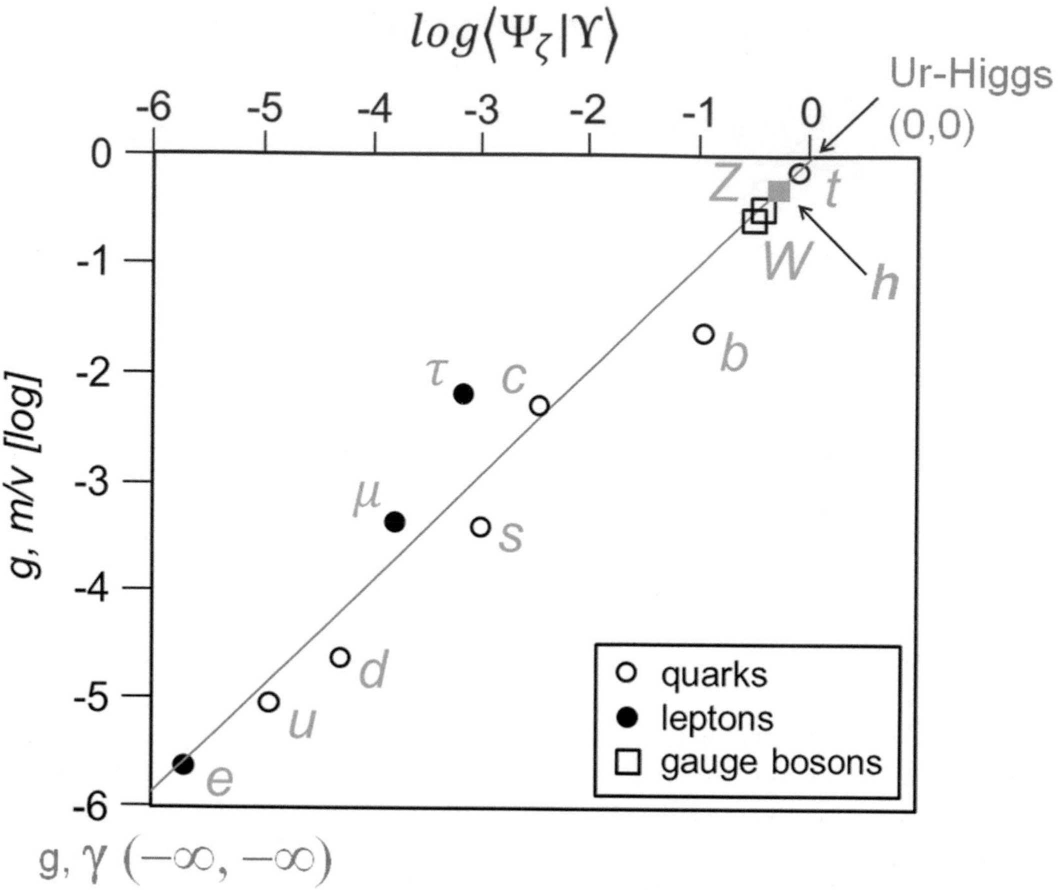

FIGURE 6.7 Prediction of the mass for each elementary particle (ζ) in the Standard Model through the correlation $\frac{m_\zeta}{v}, g_\zeta - \langle \Psi_\zeta \mid \Upsilon \rangle$ (R^2=0.88), where $\frac{m_\zeta}{v}$ is the ratio of gauge boson mass over vacuum expectation value for the Higgs boson, and g_ζ is the Yukawa coupling parameter for the case when ζ denotes a fermion. The linear correlation validates the quintessential decoding of the Standard Model as reverse engineering. Particles are denoted according to the convention described in Figure 6.4b.

ur-field. The implication is that geometric dilution, or equivalently, the pitch angle α, is a fundamental parameter that determines the mass of ur-particles originally enshrined in the ur-Higgs. When encoded into the quotient (latent) space *W*/~, this relatively simple process becomes the much more complex process by which mass is endowed by the Higgs boson in its interplay with the particle fields [2]. The commutativity of the following diagram represents the decoding of the Higgs mass-endowment process:

$$W \xrightarrow{\pi} W/\sim$$

$$\text{geometric dilution} \downarrow \widetilde{\mathfrak{F}_H} \quad \downarrow \mathfrak{F}_H \ \text{Higgs mechanism}$$

$$W \xrightarrow{\pi} W/\sim \tag{6.12}$$

The flow $\mathfrak{F}_H$ represents the endowment of mass through interaction with the Higgs boson. This process should be interpreted in the sense that for a generic particle with Lagrangian $\mathcal{L}$, the following relation holds: $\Gamma_{\mathcal{L}_H}\mathfrak{A}(\mathcal{L}) = \mathfrak{A}(\mathcal{L}')$, where the massless original Lagrangian $\mathcal{L}$ has changed to $\mathcal{L}'$, now endowed with mass conferred through interaction with the Higgs boson, while the latter becomes a Nambu-Goldstone massless boson [2]. This process is decoded at the W-level by the flow $\widetilde{\mathfrak{F}_H}$, representing the endowment of mass on the ur-particle through geometric dilution of the ur-Higgs boson (Figure 6.6). In other words, the following relation holds at the encoded W-level: $\Gamma_{\widetilde{\mathcal{L}_H}}\tilde{\mathfrak{A}}(\tilde{\mathcal{L}}) = \tilde{\mathfrak{A}}(\widetilde{\mathcal{L}'})$, where $\widetilde{\mathcal{L}_H}$ is the Lagrangian for the ur-Higgs boson, and $\tilde{\mathcal{L}}$, $\widetilde{\mathcal{L}'}$ are the decoded Lagrangians for the massless ur-particle and for the ur-particle endowed with mass through the geometric dilution determined by the operator $\Gamma_{\widetilde{\mathcal{L}_H}}$, as illustrated in the Feynmann diagram of Figure 6.6.

6.4 QUINTESSENTIAL GEOMETRIC DILUTION STEERS THE UNIVERSE'S EVOLUTION AFTER THE BIG BANG

A straightforward application of Heisenberg's uncertainty principle reveals that the emergence of the ur-Higgs, the primeval ur-particle implicated in baryogenesis, can be traced to a very specific time in the evolution of the universe after the Big Bang: $\tau \approx r_0/2c = 1.337\times10^{-27}\,s$. This takes us to the so-called electroweak epoch when electromagnetism and the weak nuclear force remained merged into the so-called electroweak force [4]. At energies in the order of the vacuum expectation energy for the Higgs field (246GeV), this merging remained energetically above the critical bound required for symmetry breaking, estimated in the SM at 159.5GeV.

The flattening of the universe begins with the expansion driven by quintessential geometric dilution with $\alpha\neq0$ (Figure 6.8), so gravity becomes less of an attribute of intrinsic geometry with extreme curvature [5] and more an attribute of the energy stored as mass in geometrically diluted incarnations of the ur-Higgs. The timing of universe evolution is in fact dictated by geometric dilution through the quantum relation

$$\tau \approx \frac{r_0}{2c}\left(\frac{1}{\cos\alpha}\right) = 1.337\times10^{-27}\left(\frac{1}{\cos\alpha}\right)s \qquad (6.13)$$

The relation (6.13) marks the timing of the creation of an ur-particle with mass $m = \hbar\cos\alpha/cr_0$.

Thus, the emergence of the decoded Higgs boson (Figure 6.9) can be traced to the time

$$\tau \approx 1.337\times10^{-27}\left(\frac{1}{\cos\alpha^*}\right)s = 2.632\times10^{-27}\,s \qquad (6.14)$$

This time marks the end of the electroweak epoch and is followed by the emergence of the decoded gauge bosons Z and W, which signal the splitting of the electroweak force, as these bosons communicate the weak nuclear force when encoded in the latent space-time

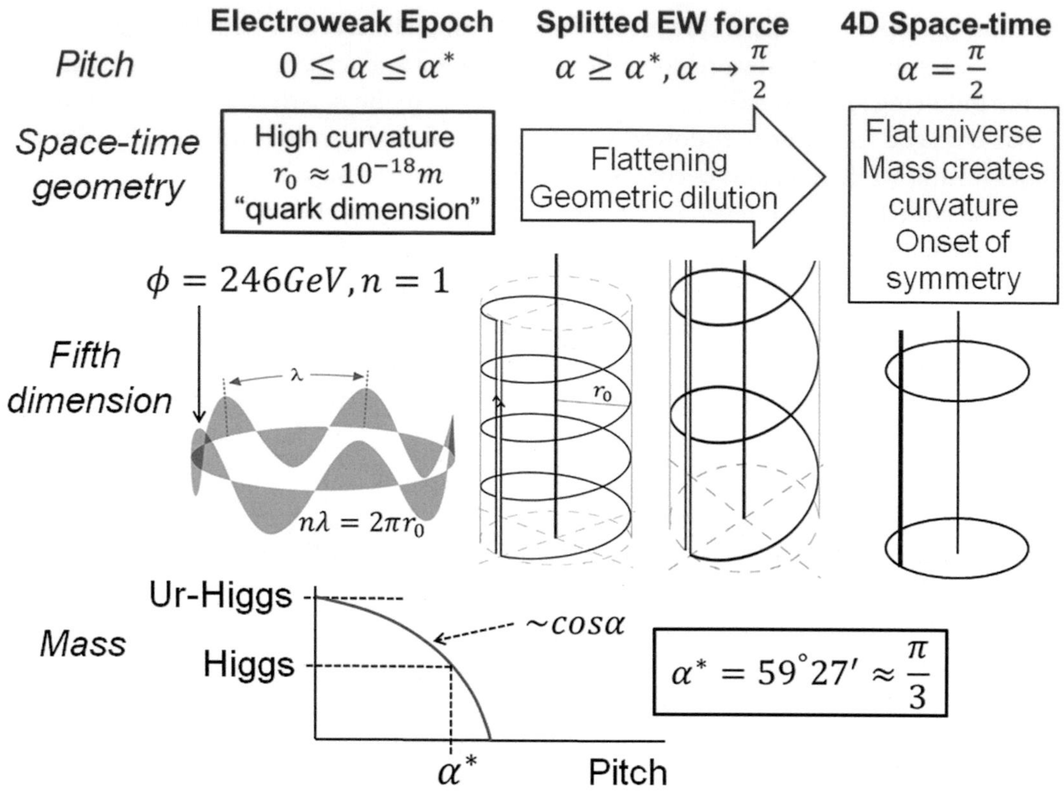

FIGURE 6.8 Flattening of the universe as a measure of time starting at the Big Bang event. The universe expansion is driven by quintessential geometric dilution with $\alpha>0$, so gravity becomes less of an attribute of intrinsic geometry of the quintessential space and more an attribute of the energy stored as mass in geometric dilutions of the ur-Higgs boson. The timing t in universe evolution starting at the Big Bang becomes $t \approx \frac{r_0}{2c}\left(\frac{1}{\cos\alpha}\right)$.

manifold $W/\sim$. All gauge bosons are created through vacuum fluctuation and stored in the diluted dormant dimension at dilution with $\cos\alpha=0.32$ (Figure 6.7), which corresponds to the electroweak splitting time $\tau \approx 4.178\times10^{-27}\,s$. The ur-fermions that emerge after this time can decay, as the gauge bosons have already stepped into existence and hence convey the interaction required (Figure 6.9).

At this point, the quintessential space W is still endowed with significant local curvature and will become flat only asymptotically at infinite geometric dilution ($\upsilon\to\infty$) corresponding to the limit $\alpha \to \pi/2$. This is the limit for the emergence of the photon (γ), the newest and massless particle. Thus, the advent of the photon becomes indicative of a quasi-flat universe resulting from infinite geometric dilution of the dormant dimension (Figure 6.9).

It is an undisputed fact that the photon already exists in our universe, and on the other hand, the Big Bang happened approximately 13.8 billion years ago, therefore the flatness of the universe should be considered to be only approximate, but it is indeed a strikingly

$$\Delta E \cdot \Delta t \sim \frac{\hbar}{2} \qquad r_0 = 0.802 \times 10^{-18}m \; (*) \qquad \phi_{vev} = \frac{\hbar c}{r_0} = 246 GeV$$

$$t \sim \left(\frac{r_0}{2c}\right)\frac{1}{\cos\alpha_{max}} = 1.337 \times 10^{-27}\left(\frac{1}{\cos\alpha_{max}}\right)s$$

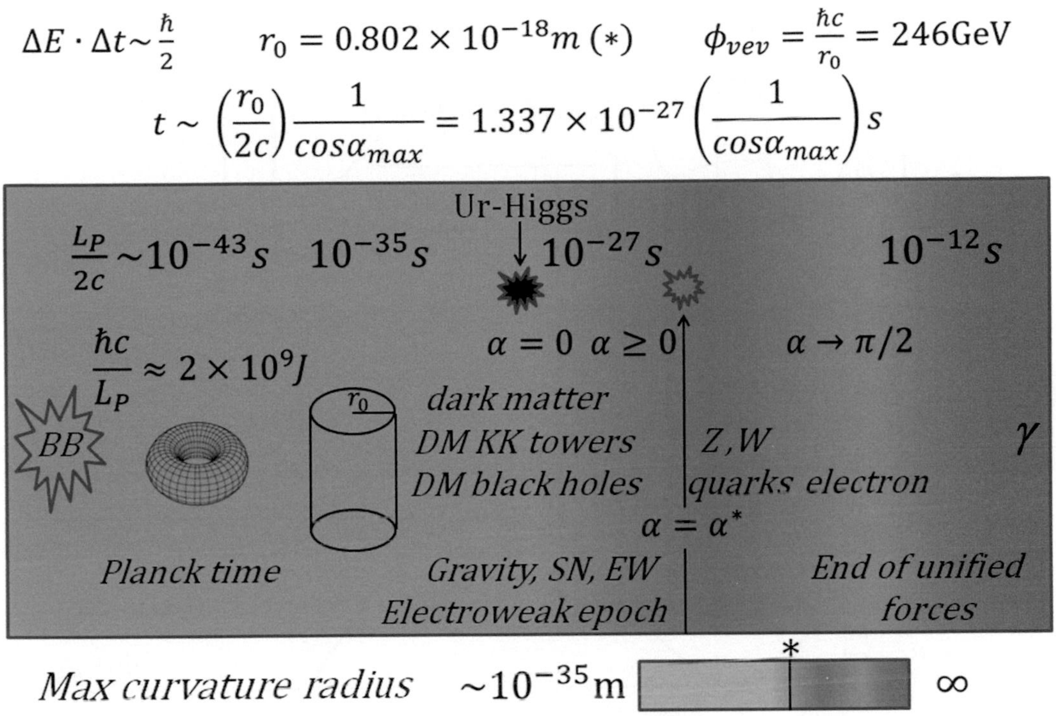

FIGURE 6.9   Universe timeline as quintessential geometric dilution.

good approximation. The current pitch $\alpha = \alpha(\tau_u)$, corresponding to the age of the universe estimated as $\tau_u \sim 4.35 \times 10^{17}s$ can be calculated effectively as

$$\alpha = \arccos\left(\frac{r_0}{2c\tau_u}\right) = \arccos\left(3.1 \times 10^{-45}\right) \cong \arccos 0 = \frac{\pi}{2} \qquad (6.15)$$

It should be noted also that the ur-photon and the photon itself constitute for all purposes one and the same entity since there is no projection onto the dormant dimension. Thus, the present day represents a singularity in which the dormant dimension may be set to be zero, and therefore, quotient space merges and identifies with underlying quintessential space. But this implies that the vacuum expectation value ϕ_{vev} for the Higgs field ($\sim \hbar c / r_0$) will ultimately be infinite and that the current value at 246Gev that sustains the universe as we know it represents a metastable state. We do not know when the current ϕ_{vev} for the Higgs field will decay to infinity, as that would depend on the energy barrier that must be overcome through quantum tunneling. But we can be certain that the universe, as we know it, is metastable [4, 5], hence prone to undergo a phase transition at some point (Figure 6.10). When that phase transition materializes, a big crunch will take place, as every particle will be endowed with infinite mass, taking the universe back to the starting point in a Big Bang scenario.

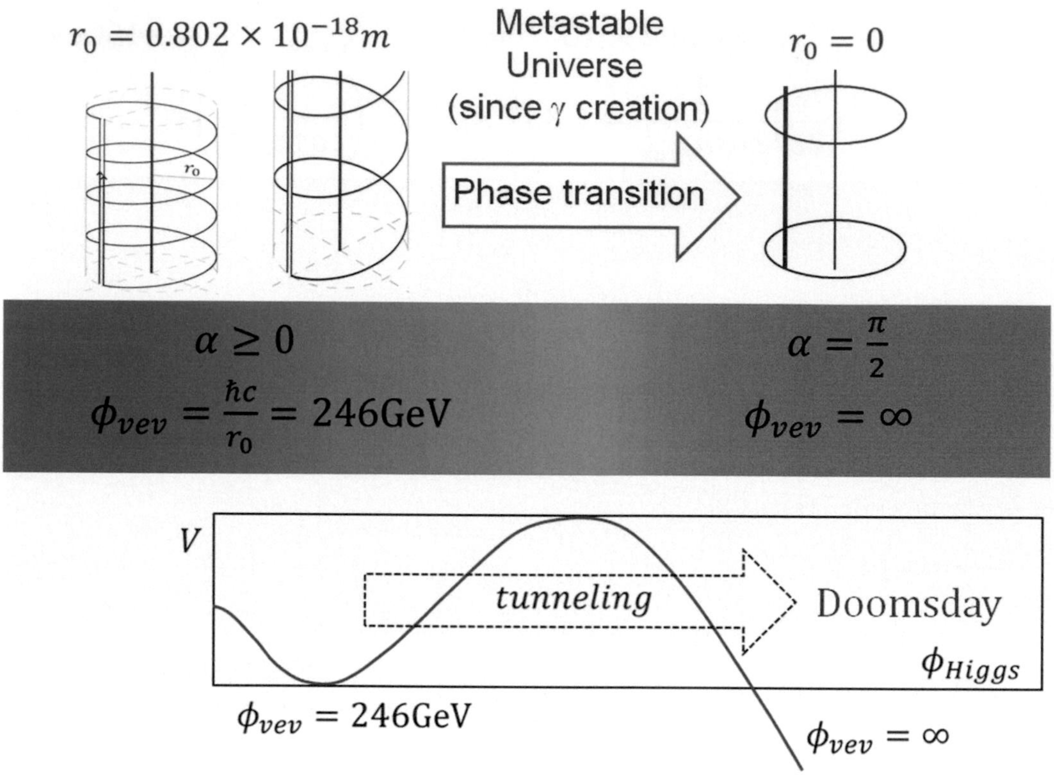

FIGURE 6.10 Catastrophe scenario associated with the metastability of the Higgs boson in the present-day universe at nearly infinite geometric dilution.

6.5 DARK MATTER AND DARK ENERGY IDENTIFIED BY THE QUINTESSENTIAL AUTOENCODER OF THE STANDARD MODEL

The conclusions drawn so far from the discussion in the preceding sections may be summarized as follows:

The universe topology is "revealed" by the CMB fluctuation spectrum, and it is identified with a four-dimensional torus, corresponding to a quasi-flat multiply connected compact space.

This topology is compatible with the Big Bang scenario for universe evolution [4], which admits dimensional expansion with retention of compactness but not a change in topology, as the latter is forbidden by general relativity [5].

The toroidal topology admits a circular dormant dimension as a compact extra dimension, whereas a Euclidean space cannot incorporate this compact dimension without a change in topology along the universe's evolution.

AI can trace the origin of DM in the universe evolution, since on a compact extra dimension with quark-size radius q [3], a de Broglie wave stores energy equal to the vacuum expectation value of the Higgs field.

This is the ur-Higgs particle, an excitation of the quintessence field that has been identified as DM [6] by AI because:

- The incorporation of a fifth dimension is supported by evidence: Elementary particle masses can be accurately predicted through the geometric dilution of the ur-Higgs quintessence field that is determined from the decoding of the particle fields (Figure 6.7).
- The ur-Higgs particle formed during the electroweak epoch in an early universe ($\sim 10^{-27}$s after the Big Bang).
- It is massive (246GeV/c^2) and cold (speed<<c).
- It does not decay through communication with SM gauge bosons.
- It is only interactive with the SM via gravity.
- It endows other ur-particles with mass through ur-boson-conveyed geometric dilution.

Much has been written about the difficulty in identifying the physical underpinnings of dark energy (DE) as the energy associated with quantum vacuum fluctuations (particles popping in and out of existence) [3]. In principle, the DE density required to observe the measured universe expansion is estimated at $\rho_{DE} \approx 5\times10^{-10}\,Jm^{-3}$. This value is in stark disagreement with naïve calculations of the vacuum energy density, yielding an estimated DE density with a colossal discrepancy amounting to a staggering 120 orders of magnitude compared with the experimentally obtained figure. In cosmology, this problem is often referred to as the cosmological constant problem or vacuum catastrophe [5]. The so-called cosmological constant [4] is often regarded as the default model for DE, whereby the geodesic fabric of space-time has a constant non-zero energy density that yields an antigravitational pull background [5]. The problem essentially describes the disagreement between the observed values of vacuum energy density (the small value of the cosmological constant) and the theoretically large value of zero-point energy that is obtained from quantum field theory. As said, the quantum vacuum energy contribution to the cosmological constant is calculated to be as much as 120 orders of magnitude greater than the one observed, a calamitous state of affairs that should in all likelihood be referred to as the largest disagreement between theory and experiment the history of physics. This prompted scientists to explore other options for what DE might be.

Much of the difficulty and controversy evaporates as the extra compact dimension of quark-like material scale is incorporated in the quintessential decoding of the latent manifold representing standard space-time. This is because the excitation of the quintessential ur-Higgs field through quantum vacuum fluctuations enables the storage of DE as DM. In other words, the AI-enabled autoencoder technology dictates that we cannot conceive DE independently of DM and the dichotomy becomes a consequence of the quintessential decoding of our universe through the incorporation of the dormant fifth dimension.

To distill the thrust of this discussion, we may state that the decoding of the SM by a quintessential autoencoder yields an ur-field that is not part of the SM. This quintessential field serves as the excitation vehicle for DE. In turn, the DE gets occasionally stored as DM for wavelengths that fit a stationary wave condition along the dormant fifth dimension. Thus, AI shows that DE and DM become phenomenologically tied up to the quintessential space while becoming opaque to our understanding when regarded within the standard four-dimensional space-time. Only when the latter is regarded as a latent space, the true nature of DM and DE may be revealed.

6.6 GENERATING DARK MATTER AND DARK ENERGY IN THE HOLOGRAPHIC AUTOENCODER WITH GEOMETRIC DILUTION AS A PROXY FOR TIME

In Chapter 5, the challenging problem of quantum gravity was addressed by constructing a learning system with stochastic connectivities and hidden variables where gravity and quantum behavior become emergent properties in a statistical physics scheme for machine learning. The thrust was to implement an AI-based version of the universe as a quantum holographic autoencoder. This system operates under the tenet that the emergent quantum behavior arises in a neural network equilibrated on the nontrainable hidden variables upon which a relativistic string gravitational scheme may be constructed. Conversely, in the nonequilibrium regime prior to the equilibration of nontrainable variables, the network is endowed with emergent gravity. The behavior of the neural network is examined in the limits where the bias vector, weight matrix, and state vector of neurons can be modeled as stochastic variables that undergo a learning evolution. These dynamics are described by a time-dependent Schrödinger equation, Equation (5.40), compatible with the relativistic decoding enshrined in the commutativity of the diagram in Equation (5.44). The results illustrate the emergence of quantum mechanics and general relativity in neural networks governed by a statistical physics scheme that can operate in two different thermodynamic regimes. In this way, the quantum metamodel for gravity fulfills at least in part a major imperative for physicists seeking a unified field theory.

The holographic autoencoder is now set up to adopt the toroidal topology of space-time described in Figure 6.3 with the geometric dilution parameter $\upsilon = -\log\cos\alpha$ as the proxy for time. With the default parameters given in Chapter 5, the timing of events follows exactly the description in Figure 6.9, with the advent of dark matter at ~10^{-27}s. The Lagrangian for the universe evolves as described in Figure 6.11 in parametric dependence with the geometric dilution parameter. As dictated by the holographic autoencoder, the complete materialization of the universe and its evolution within the Big Bang scenario can be realized through the geometric dilution of gravity, originally decoded as a stationary wave in the quintessential five-dimensional torus.

6.7 THE UNIVERSE IN THE QUINTESSENTIAL REVERSE AUTOENCODER: NO EXTENSION OF THE STANDARD MODEL MAY YIELD DARK ENERGY OR DARK MATTER

The concept of reverse autoencoder, or rather, "autodecoder," has featured profusely in the previous discussion on the AI-enabled reverse engineering of the SM. In time-series data science, the autoencoder is regarded as the time-series distiller that generates the essential

(a)

0
$\cos\alpha = 1$
Ur – Higgs

1
$\cos\alpha \geq 0.703$
gluons, top quark

1+2
$\cos\alpha \geq 0.32$
gluons, top quark, gauge bosons

1

$$-\tfrac{1}{2}\partial_\nu g^a_\mu \partial_\nu g^a_\mu - g_s f^{abc}\partial_\mu g^a_\nu g^b_\mu g^c_\nu - \tfrac{1}{4} g_s^2 f^{abc} f^{ade} g^b_\mu g^c_\nu g^d_\mu g^e_\nu + \tfrac{1}{2} i g_s^2 (\bar{q}^\sigma_i \gamma^\mu q^\sigma_j) g^a_\mu + \bar{G}^a \partial^2 G^a + g_s f^{abc} \partial_\mu \bar{G}^a G^b g^c_\mu$$

2

$$- \partial_\nu W^+_\mu \partial_\nu W^-_\mu - M^2 W^+_\mu W^-_\mu - \tfrac{1}{2}\partial_\nu Z^0_\mu \partial_\nu Z^0_\mu - \tfrac{1}{2c_w^2} M^2 Z^0_\mu Z^0_\mu - \tfrac{1}{2}\partial_\mu A_\nu \partial_\mu A_\nu - \tfrac{1}{2}\partial_\mu H \partial_\mu H - \tfrac{1}{2} m_h^2 H^2 - \partial_\mu \phi^+ \partial_\mu \phi^- - M^2 \phi^+ \phi^- - \tfrac{1}{2}\partial_\mu \phi^0 \partial_\mu \phi^0 - \tfrac{1}{2c_w^2} M \phi^0 \phi^0 - \beta_h [\tfrac{2M^2}{g^2} + \tfrac{2M}{g} H + \tfrac{1}{2}(H^2 + \phi^0\phi^0 + 2\phi^+\phi^-)] + \tfrac{2M^4}{g^2}\alpha_h - igc_w[\partial_\nu Z^0_\mu (W^+_\mu W^-_\nu - W^+_\nu W^-_\mu) - Z^0_\nu (W^+_\mu \partial_\nu W^-_\mu - W^-_\mu \partial_\nu W^+_\mu) + Z^0_\mu (W^+_\nu \partial_\nu W^-_\mu - W^-_\nu \partial_\nu W^+_\mu)] - igs_w[\partial_\nu A_\mu (W^+_\mu W^-_\nu - W^+_\nu W^-_\mu) - A_\nu (W^+_\mu \partial_\nu W^-_\mu - W^-_\mu \partial_\nu W^+_\mu) + A_\mu (W^+_\nu \partial_\nu W^-_\mu - W^-_\nu \partial_\nu W^+_\mu)] - \tfrac{1}{2} g^2 W^+_\mu W^-_\mu W^+_\nu W^-_\nu + \tfrac{1}{2} g^2 W^+_\mu W^-_\nu W^+_\mu W^-_\nu + g^2 c_w^2 (Z^0_\mu W^+_\mu Z^0_\nu W^-_\nu - Z^0_\mu Z^0_\mu W^+_\nu W^-_\nu) + g^2 s_w^2 (A_\mu W^+_\mu A_\nu W^-_\nu - A_\mu A_\mu W^+_\nu W^-_\nu) + g^2 s_w c_w [A_\mu Z^0_\nu (W^+_\mu W^-_\nu - W^+_\nu W^-_\mu) - 2A_\mu Z^0_\mu W^+_\nu W^-_\nu] - g\alpha[H^3 + H\phi^0\phi^0 + 2H\phi^+\phi^-] - \tfrac{1}{8} g^2 \alpha_h [H^4 + (\phi^0)^4 + 4(\phi^+\phi^-)^2 + 4(\phi^0)^2\phi^+\phi^- + 4H^2\phi^+\phi^- + 2(\phi^0)^2 H^2] - gMW^+_\mu W^-_\mu H - \tfrac{1}{2} g \tfrac{M}{c_w^2} Z^0_\mu Z^0_\mu H - \tfrac{1}{2} ig[W^+_\mu (\phi^0 \partial_\mu \phi^- - \phi^- \partial_\mu \phi^0) - W^-_\mu (\phi^0 \partial_\mu \phi^+ - \phi^+ \partial_\mu \phi^0)] + \tfrac{1}{2} g [W^+_\mu (H \partial_\mu \phi^- - \phi^- \partial_\mu H) - W^-_\mu (H \partial_\mu \phi^+ - \phi^+ \partial_\mu H)] + \tfrac{1}{2} g \tfrac{1}{c_w} (Z^0_\mu (H \partial_\mu \phi^0 - \phi^0 \partial_\mu H) - ig \tfrac{s_w^2}{c_w} M Z^0_\mu (W^+_\mu \phi^- - W^-_\mu \phi^+) + igs_w M A_\mu (W^+_\mu \phi^- - W^-_\mu \phi^+) - ig \tfrac{1-2c_w^2}{2c_w} Z^0_\mu (\phi^+ \partial_\mu \phi^- - \phi^- \partial_\mu \phi^+) + igs_w A_\mu (\phi^+ \partial_\mu \phi^- - \phi^- \partial_\mu \phi^+) - \tfrac{1}{4} g^2 W^+_\mu W^-_\mu [H^2 + (\phi^0)^2 + 2\phi^+\phi^-] - \tfrac{1}{4} g^2 \tfrac{1}{c_w^2} Z^0_\mu Z^0_\mu [H^2 + (\phi^0)^2 + 2(2s_w^2 - 1)^2 \phi^+\phi^-] - \tfrac{1}{2} g^2 \tfrac{s_w^2}{c_w} Z^0_\mu \phi^0 (W^+_\mu \phi^- + W^-_\mu \phi^+) - \tfrac{1}{2} ig^2 \tfrac{s_w^2}{c_w} Z^0_\mu H (W^+_\mu \phi^- - W^-_\mu \phi^+) + \tfrac{1}{2} g^2 s_w A_\mu \phi^0 (W^+_\mu \phi^- + W^-_\mu \phi^+) + \tfrac{1}{2} ig^2 s_w A_\mu H (W^+_\mu \phi^- - W^-_\mu \phi^+) - g^2 \tfrac{s_w}{c_w} (2c_w^2 - 1) Z^0_\mu A_\mu \phi^+ \phi^- - g^1 s_w^2 A_\mu A_\mu \phi^+ \phi^-$$

(b)

1+2+3 $\cos\alpha \geq 2.07 \times 10^{-6}$

Elementary particles interact with gauge bosons that communicate the weak force and with the Higgs field. Weak force enables decay of massive particles into lighter particles.

3

$$\bar{d}^\lambda_j (\gamma\partial + m^\lambda_d) d^\lambda_j + igs_w A_\mu [-(\bar{e}^\lambda \gamma^\mu e^\lambda) + \tfrac{2}{3}(\bar{u}^\lambda_j \gamma^\mu u^\lambda_j) - \tfrac{1}{3}(\bar{d}^\lambda_j \gamma^\mu d^\lambda_j)] + \tfrac{ig}{4c_w} Z^0_\mu [(\bar{\nu}^\lambda \gamma^\mu (1+\gamma^5)\nu^\lambda) + (\bar{e}^\lambda \gamma^\mu (4s_w^2 - 1 - \gamma^5) e^\lambda) + (\bar{u}^\lambda_j \gamma^\mu (\tfrac{4}{3} s_w^2 - 1 - \gamma^5) u^\lambda_j) + (\bar{d}^\lambda_j \gamma^\mu (1 - \tfrac{8}{3} s_w^2 - \gamma^5) d^\lambda_j)] + \tfrac{ig}{2\sqrt{2}} W^+_\mu [(\bar{\nu}^\lambda \gamma^\mu (1+\gamma^5) e^\lambda) + (\bar{u}^\lambda_j \gamma^\mu (1+\gamma^5) C_{\lambda\kappa} d^\kappa_j)] + \tfrac{ig}{2\sqrt{2}} W^-_\mu [(\bar{e}^\lambda \gamma^\mu (1+\gamma^5)\nu^\lambda) + (\bar{d}^\kappa_j C^\dagger_{\lambda\kappa} \gamma^\mu (1+\gamma^5) u^\lambda_j)] - \bar{e}^\lambda (\gamma\partial + m^\lambda_e) e^\lambda - \bar{\nu}^\lambda \gamma\partial \nu^\lambda - \bar{u}^\lambda_j (\gamma\partial + m^\lambda_u) u^\lambda_j$$

FIGURE 6.11 (a, b) Evolution of the Lagrangian of the universe as generated by the holographic autoencoder of quantum gravity with geometric dilution as the proxy for time. The notation by T. A. Gutierrez is universally adopted for the Standard Model of elementary particle physic and has been deduced from Veltman M (1994) *Diagrammatica: The Path to Feynmann Diagrams.* Cambridge University Press, Cambridge, UK.

or latent model in a dynamical system [1]. In our context of interest, we assume the SM is, in and of itself, the latent model that entrains the full dynamics, and the leveraging of reverse autoencoder technology serves the purpose of determining the latter.

As a specialized learning system, the autoencoder typically needs to be trained to decode the latent dynamics. In the particular case of interest, the training does not require *a priori* knowledge of quintessential time series. This is because we have a generic way of deriving the quintessential Lagrangian $\tilde{\mathcal{L}}$ that decodes a particle Lagrangian $\mathcal{L}$. Thus, the decoded Lagrangian satisfies the relation: $\int \tilde{\mathcal{L}}(\tilde{\phi})\, d\boldsymbol{x} dy_5 = \int \mathcal{L}(\phi)\, d\boldsymbol{x} + [\int |\int(\tilde{\phi} - \phi) d\boldsymbol{x}|^2\, dy_5]^{1/2}$. By applying this relation to a given pair $(\phi, \mathcal{L})$ representing a particle in the SM, the dynamics encoded in the SM are fleshed out onto the quintessential space (a compact and locally flat multiply connected manifold). This simply requires decoding each particle field $\phi \rightarrow \tilde{\phi}$ so as to optimize (minimize) the term $\int \tilde{\mathcal{L}}(\tilde{\phi})\, d\boldsymbol{x} dy_5$.

To deploy autoencoder technology, we have turned the SM into a dynamical system by setting the time step for an elementary process equal to the lifetime of the (ephemeral) boson that mediates the particle transformation. Evidently, the center manifold entrainment [7–9] or subordination of the quintessential dynamical system to the SM leaves out dark energy and dark matter, so we cannot assume that the SM, staggeringly successful as it is, provides the complete description of the elementary processes that take place in the universe.

Our discussion and analysis in this chapter have focused primarily on elucidating what is left out with the reduction or entrainment of the quintessential dynamics by the SM, and we have found a striking answer: DM and DE. This is at some level disconcerting because both clearly have a very significant bearing on the observed dynamics of deep space. To be specific, they constitute over 95% of the gravitational budget of the detectable universe at large scales. Furthermore, this clearly introduces a contradiction: The existence of DM and DE implies that the SM cannot entrain or subordinate the universe as a dynamical system that materializes in the quintessential space. This contradiction can only be properly accounted for in one of two ways: (1) The SM does not represent the latent dynamics of the universe, or (2) the quintessential dynamics are in fact irreducible: They do not admit a latent dynamical system. This leads to a paradox since any of the two alternatives implies that an autoencoder could not have been *a priori* used to elucidate the quintessential dynamics spanned by the latent dynamics enshrined in the SM. We know this to be wrong since an autoencoder has indeed been used as a decoder of the SM.

This fundamental paradox does not undermine the AI enablement of the reverse engineering of the SM because it is based on the obviously incorrect assumption that DM and DE can be obtained from extensions of the SM. This means that DM and DE are inherently quintessential, which was from the start a basic tenet in the AI approach put forth in this book. The fundamental implication is that the decoding of the SM is insufficient to account for the quintessence fields that realize DM and DE when excited along the dormant dimension.

Notwithstanding the previous conclusions, experimental efforts engaging multinational consortia, like the Large Hadron Collider, or deploying underground detectors are still well underway to detect DE and DM as extensions of the SM. Such efforts have been, not surprisingly, unsuccessful. It seems that for all that we revere the Copernican revolution, we still like to see ourselves as the center of the universe, and AI may be teaching us a lesson in that regard...

REFERENCES

1. Fernández A (2022) *Topological Dynamics for Metamodel Discovery with Artificial Intelligence.* Chapman & Hall/CRC, Taylor and Francis, London
2. Feynman RP, Weinberg S (1999) *Elementary Particles and the Laws of Physics.* Cambridge University Press, Cambridge, UK
3. Abramowiczy H, Abtt I, Adamczykh L, Adamusae M, Antonelli S, et al., Zeus Collaboration (2016) Limits on the effective quark radius from inclusive *ep* scattering at HERA. *Phys Lett B* 757: 468–472
4. Weinberg S (2008) *Cosmology.* Oxford University Press, New York
5. Hawking SW, Ellis GFR (2023) *The Large Scale Structure of Space-Time: 50th Anniversary Edition.* Cambridge University Press, Cambridge, UK
6. Profumo S (2017) *Introduction to Particle Dark Matter.* World Scientific Publishing, Singapore
7. Fernández A (1985) Center-manifold extension of the adiabatic-elimination method. *Phys Rev A* 32: 3070–3076
8. Fernández A (1988) Phase-ordering dynamics for the onset of a center manifold. *Phys Rev A* 38: 4256–4262
9. Fernández A (1988) On renormalization of fluctuations at the onset of a centre manifold. *J Phys A: Math Gen* 21: L607

Epilogue

Conversion of Dark Energy into Dark Matter with Cosmic Reproduction Technology

It seems probable that once machine thinking had started,
it would not take long to outstrip our feeble powers.

– ALAN TURING

E.1 THE COSMOLOGICAL CONSTANT PROBLEM IN A MULTIVERSE MATRIX

AI has addressed the quantum gravity problem in more than one way, and one alternative is constructing a "physical learning machine" with stochastic connectivity weights and hidden variables representing the random states of the nodes [1]. This system may be endowed with emergent gravity and emergent quantum behavior. In this way, AI represents the universe as a holographic autoencoder [1], where the holographic map h: $\partial W \rightarrow W$ is one of the many possible inverses of the canonical projection π: $W \rightarrow \partial W$ that maps the quintessential supra-relativistic torus W onto its "quantum border" ∂W. The emergent quantum behavior arises in the equilibrium regime for node state (hidden) variables upon which relativistic strings are embroidered [2]. Conversely, the network is endowed with an emergent gravity in the nonequilibrium regime for the hidden variables. The results illustrate the possibility of establishing the duality of quantum mechanics/general relativity in neural networks regarded as physical systems governed by the laws of statistical mechanics (Chapter 5). These networks are endowed with two different thermodynamic regimes, where the geometric dilution parameter $\upsilon = -\log\cos\alpha$ is the proxy for time, and the advent of the lightest dark matter occurs at $\sim 10^{-27}$s counting from the birth of the universe.

A birth channel for cosmic reproduction can be set up within this scheme by constructing a second holographic autoencoder $h' : \partial W' \rightarrow W'$ that serves as an antenna, capable of capturing the tunneling $\partial W \rightarrow \partial W'$ of a dark energy burst in the universe (∂W, W) in the form of a quantum vacuum fluctuation. The tunneled fluctuation is amplified within the second holographic autoencoder to give birth to a progeny universe $(\partial W', W')$. By $\sim 10^{-27}s$ after its birth, the baby universe will begin storing the tunneled dark energy in a material

DOI: 10.1201/9781003385950-7

embodiment, with the formation of dark matter in the form of ur-Higgs particles. In this way, AI constructs a cosmic reproductive machine that converts dark energy into dark matter (Figure E.1). Notably, this machine harnesses dark energy as the cosmic "birth inducer" and generates dark matter in a multiverse scenario as the canonical (surjective but not injective) projection $\pi' : W' \rightarrow \partial W'$ admits *a priori* a multiplicity of holographic maps $h' : \partial W' \rightarrow W'$ all satisfying the equation $h' \circ \pi' = id_{W'}$.

To leverage this multiverse scenario, an advanced AI-based cosmic technology for universe reproduction is implemented. The technology harnesses dark energy tunneling as a nucleating quantum fluctuation that spans the progeny universe. The cosmic reproduction machine requires a second holographic autoencoder serving as an antenna or $\partial W \rightarrow \partial W'$ receiver. The enormous energy surplus from vacuum quantum energy that is meant to be

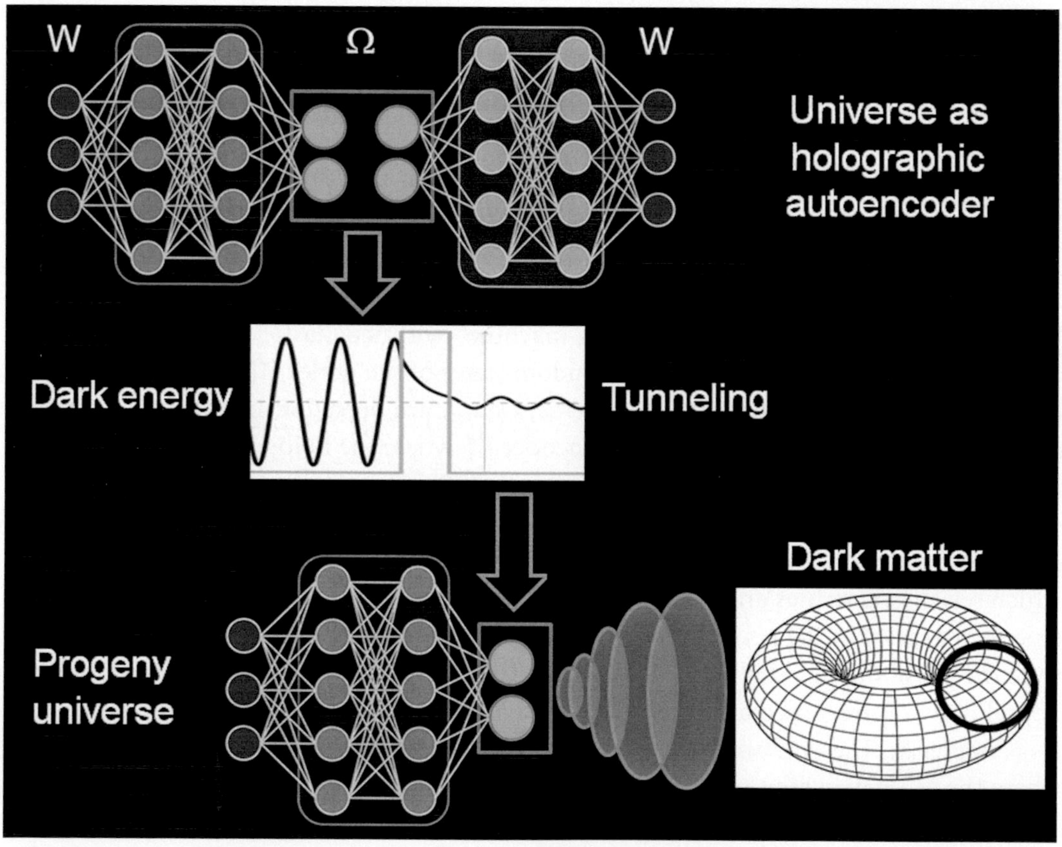

FIGURE E.1 An AI-based cosmic reproductive machine that converts dark energy into dark matter in a multiverse matrix. A birth canal for cosmic reproduction is set up by coupling the universe ($\Omega = \partial W, W$) to a second holographic autoencoder acting as receiver and capable of capturing the tunneled dark energy originating in the progenitor universe. The tunneled quantum fluctuation is amplified through the receiver autoencoder to give birth to the progeny universe ($\Omega' = \partial W', W'$), which stores the tunneled dark energy as dark matter in the form of stationary waves along the quintessential circular coordinate of the quark-size attometer dimension.

indicative of a poor understanding of dark energy in our universe may be thus reconciled within the multiverse scenario put together by a technologically advanced civilization capable of cosmic investment in dark energy for reproductive purposes, as schematically depicted in Figure E.1.

The vacuum catastrophe or cosmological constant problem (Chapter 3) alluded to above is arguably the most embarrassing discrepancy in all of physics. It has become a veritable "tag for demolition" of the whole edifice of physics. A naïve computation of the vacuum quantum energy density yields values estimated to be 120 orders of magnitude larger than the expected contribution to the cosmological constant based on experimental observation [3]. AI provides a solution to this problem, shifting the task to a mere calculation of the rate of generation of progeny universes arising from dark energy tunneling. Thus, the tunneling of vacuum fluctuations is given a material embodiment as dark matter ur-Higgs particles in the embryonic universe. These particles are stored as stable wave excitations along the quintessential quark-scale dimension and endowed with mass $246\mathrm{GeV}/c^2$, the vacuum expectation value of the Higgs boson.

This is obviously AI's multiverse solution to the cosmological constant problem. The solution is based on a "multiverse matrix," a pivotal scenario consisting of coupled holographic autoencoders capable of converting dark energy into dark matter as the tunneling of dark energy is steered through a cosmic birth canal.

E.2 RELATIVITY MADE "PHYSICAL" – THE IMPRINT OF THE UNIVERSE-SIMULATING PROCESSOR

Far-fetched as it may seem, the hypothesis that the universe is a simulation, put forth by the philosopher Nick Bostrom [4], has been assigned a staggering 50:50 chance of being true. It has also been the object of derision, as it has been deemed to be a non-scientific idea. This is because, it has been argued, it is incapable of yielding a falsifiable prediction. The idea, today attributed to a philosopher, once belonged *mutatis mutandis* to a literary province. Translate the computer science term "simulation" as "dream" and we get Bostrom's hypothesis in a literary embodiment, as illustrated by the story of the butterfly dreamed by Zhuangzi, the play "La Vida es Sueño" (Life is a Dream) by Pedro Calderón de la Barca, or the short story "The Circular Ruins" by Jorge Luis Borges.

Be that as it may, the simulation hypothesis has recently met a major scientific standard with flying colors: It has materialized as a physical model of the universe known as a holographic autoencoder, a computational system capable of reconciling general relativity and quantum mechanics [1]. This is no minor feat since the respective material scales to which those theories apply are incommensurable, and hence the theories were deemed irreconcilable.

Any computer simulation has embossed in it the imprint of the processor responsible for its generation. In our case, this imprint should surface in the knowable universe as an "anomaly." However, our awareness of this anomaly is numbed, clouded by the extreme familiarity with our universe acquired through our conscience, a familiarity sculpted by the accretion of all our experiences than begin in the cradle. Evidently, the civilization that is simulating our universe and constructing the holographic autoencoder is inconceivably more advanced than ours, if nothing else because it has shown to be capable of cosmic manipulation (and we are not). Yet, regardless of the degree of development, the speed of the processor this civilization was able to deploy cannot be infinite. Furthermore, we know with astounding accuracy what this processor speed is! It is related to the storage capacity of the information required for the processor to perform a single operation. If the category "physical space" is being simulated and the processor performs, say, a single operation per second, then the amount of space encoded as bits of information generated must be translatable into actual physical space covering ~300,000 km along a single dimension. That is, the processor speed begets the speed of light (~300,000 km/s) in its simulated universe, making it a staggering performer, indeed!

If indeed the speed of light, an absolute upper limit in our universe, contains the imprint of the processor that simulates it, then this speed must be constant, irrespective of the observer that measures it, because it is not "physical" but an *object d'art* of the simulation. And indeed it is! In fact, this is the major tenet of relativity, and Einstein had to postulate the slowing down of time for an observer traveling at speeds close to the speed of light (relativistic speeds) in order to accommodate the artifact of constant measured speed of light: For that particular "relativistic" observer (A), the distance traveled by light is shorter than that for an observer (B) that moves at slower (nonrelativistic) speeds, but time runs more slowly for A than for B, so that the speed of light measured by both observers is the same. In other words, Einstein had to tinker with time to turn the computational artifact accurately described by the proposition "light speed is constant" into a "physical" proposition. In this sense, space-time becomes Einstein's way to circumvent, or rather come to grips with, the anomaly introduced by the fact that our universe is a simulation, something he could not anticipate at the time when his theory of relativity was conceived.

If the universe is indeed a simulation, the processor performance for maximum quark-level resolution (~1 attometer) must be a staggering $f \sim (3\times10^{8}/10^{-18})\,\mathrm{Hz} = 3\times10^{14}\,\mathrm{Teraflops}$, which is about 10^{13} times higher than the number of floating-point operations per second accessible to quantum computation. However, there are ways to circumvent the prohibitively fast performance required. This entails using autoencoders trained to distill (encode) the coarse-grained version of reality at human eye resolution, i.e. at 10^{-5}m, and decode the information all the way back to a quark-resolved reality in special cases when there is a need for such resolution (i.e., in a lab experiment broadly defined). In a quotient space resolving reality at a "human scale", the required processor speed would be $f \sim (3\times10^{8}/10^{-5})\mathrm{Hz} = 30$ Teraflops, which is commensurate with the performance of today's quantum computers.

Life may indeed be a dream.

REFERENCES

1. Fernández A (2022) *Topological Dynamics for Metamodel Discovery with Artificial Intelligence*. Chapman & Hall/CRC, Taylor & Francis, London
2. Vanchurin V (2021) Toward a theory of machine learning. *Mach Learn Sci Technol* 2: 036012
3. Adler R, Casey B, Jacob O (1995) Vacuum catastrophe: An elementary exposition of the cosmological constant problem. *Am J Phys* 63: 620–626
4. Bostrom N (2003) Are we living in a computer simulation? *Philos Quarterly* 53: 243–255

Appendix

A.1 MODEL DISCOVERY WITH AUTOENCODERS AND TRANSFORMERS: *IN VIVO* PHYSICS APPLICATIONS

This part illustrates the power of autoencoders and more specialized NN architectures to discover models and metamodels for dynamical systems of great hierarchical complexity, as encountered in molecular *in vivo* environments. Specifically, the inference of *in vivo* protein folding pathways constitutes a daunting challenge because of the molecular complexity of cellular environments. Nevertheless, it is imperative that we solve this problem because natural proteins have evolved to fold in an *in vivo* context, not in the test tube. Even for *in vitro* contexts, efforts based on molecular dynamics (MD) face great hurdles since, with few exceptions, realistic folding timescales are out of reach and rare events are missed. In more realistic settings, computational efforts at the atomistic level cannot account for the role of cellular contexts that expedite the folding process and prevent the chain from getting kinetically trapped in misfolded states. The application of artificial intelligence (AI) is likely to improve this state of affairs. We empower MD by subsuming short atomistic simulations within a "transformer" platform that propagates folding trajectories encoded in a "textual" coarse-grained representation of the chain conformation. The goal is to reverse-engineer the cooperative *in vivo* context by constructing a metamodel assisted by AI. The methods are applied to the GroEL chaperone, whose assistance significantly improves folding efficacy. In the apo state, the GroEL chamber containing the folding substrate is shown to disrupt misfolded states in a cyclic choreography of annealing iterations involving alternating subunits. In the $(ATP)_7$-state, the chaperone commits the chain to fold by buttressing nucleating native backbone hydrogen bonds that would not prevail in bulk solvent. Similarly, shedding light on the molecular processes that take place within a molecular assemblage of about 4.5 million daltons is no minor feat, and is clearly off-limits for any technology based solely on molecular dynamics. In this regard, the results obtained with AI become iconic and epitomize the power and efficacy of AI-based metamodels to handle the wanton molecular complexities of *in vivo* reality. Thus, AI is shown to play a pivotal role in unraveling the molecular complexities of cellular settings, yielding experimentally validated results.

A.1.1 Transformer Autoencoders for *In Vivo* Molecular Contexts

Due to their wanton complexity, the majority of molecular processes that take place *in vivo* have proven unyielding to computational modeling. This situation is likely to change with the leveraging of AI technologies that have already proven capable of creating and handling hierarchical representations of complex molecular information and making accurate predictions [1]. This appendix focuses on the deployment of AI to deal with the next frontier: Time-dependent molecular transformations that take place in *in vivo* settings.

Molecular dynamics (MD) is a primary tool to investigate biomolecular processes *in vitro*, such as protein folding [2, 3]. It is now believed that such efforts may have been misplaced in some instances [4–6]. For example, natural proteins actually fold and have evolved to fold in the cell, not in the test tube, with only a handful of small single-domain proteins finding their native structure *in vitro* [7–9]. With a significant proportion of proteins denaturing irreversibly *in vitro* ([6] and references therein), the Anfinsen scenario, where the protein finds its native fold *in vitro* under thermodynamic control, may require reassessment and revision, even if the Anfinsen principle asserting that structure is encoded solely by amino acid sequence remains valid [7].

Through a steering stochastic participation so far largely unidentified at the molecular level, the *in vivo* context prevents the protein from getting kinetically trapped in misfolded states and provides a cooperative context to expedite folding [4–6]. It is thus likely that the single-versus-multiple pathway controversy [10, 11] may need to be redirected to address the *in vivo* context. On its own, MD seems underequipped to reproduce such complexities at realistic catalytic turnover times. Even in the more tractable *in vitro* context, the difficulties faced by MD are apparent, as physically relevant timescales are often inaccessible, conformation space is sparsely sampled, and rare events are often missed [6, 12–14]. This appendix squarely addresses these challenges, heralding the leverage of AI technologies to empower the reverse-engineering of the *in vivo* context that accounts for the expediency of the protein folding process.

The huge informational burden that needs to be carried over from one integration step to the next makes it virtually impossible for MD to recreate an *in vivo* reality. To address the problem, AI becomes essential. Implementing an AI-empowered MD-trained system requires that we first encode or project the protein chain dynamics onto a coarse-grained representation retaining only essential topological features of the vector field that steers the integration process [9, 12]. This mathematical approach was delineated in Chapter 5 and is now specialized to the *in vivo* protein folding problem. We simplify the flow by (a) lumping conformations within basins of attraction of potential energy minima and other singularities [9], and (b) encoding the dynamics into a modulo-basin "textual" representation that effectively averages out fast motion within the basins. Inter-basin transitions in the metamodel are determined by a transformer network [15] where the receptive field for a basin transition at a residue site becomes progressively expanded within hidden layers consisting of long short-term memory (LSTM) modules parametrized by the attention range, denoted g. Subsequently, the weighted influence of different contexts upon each basin transition is variationally optimized by contrasting transformer inferences against experimentally determined structures [16]. With a suitable propagation scheme obtained by training the transformer with big dynamic data from atomistic MD runs, this AI-empowered metamodel scheme enables coverage of realistic timescales and recreates cooperative *in vivo* reality, while atomistic level detail is recovered through AI-based decoding [16].

To enable AI to construct a metamodel of *in vivo* reality and assess its role in expediting the folding process, we incorporate the influence of specific structurally reported cellular machinery. This is accomplished by training the transformer with atomistic MD simulations of folding events within specific cellular compartments. None of the MD runs can cover realistic folding timescales, but together they inform the transformer on the steering cooperative influence of the *in vivo* context.

For the sake of illustration, we adopt a training set consisting of MD runs that incorporate the influence of GroEL, a widely studied deca-tetrameric ATP-consuming chaperone [4], on protein folding [5]. Inside the GroEL chamber in the apo state, our transformer-enhanced atomistic MD runs reveal an intermittent alternating annealing process where non-compact states with a large radius of gyration (i.e., scaling with the unfolded ensemble average) interact strongly with structurally disordered hydrophobic residues in the GroEL (GGM)-enriched C-terminus tails. These interactions are cyclically choreographed with alternating subunit participation to stochastically disrupt metastable states that would constitute kinetic traps if the folding process were to take place in bulk solvent. On the other hand, when the catalytic chamber is in the $(ATP)_7$-state, it is able to boost folding expediency via enhanced intermolecular interactions that stabilize the folding-nucleating hydrogen-bond pattern of the substrate.

Thus, the influence of the *in vivo* context is incorporated through the training of the transformer in a selected environment of molecular complexity where only sub-μs events would be accessible to standard atomistic MD computation. The transformer generates a dynamics in which two fundamental aspects of chaperone assistance become apparent: (a) Coarse-grained misfolded states are levied inter-basin transitions that eventually get them dismantled, thereby preventing the protein from getting kinetically trapped; and (b) the folding-nucleating core gets protected stochastically through intermolecular interactions.

The intent of this appendix is to empower MD by leveraging AI to incorporate big dynamic data in cellular settings. Via feature extraction, our transformer reverse-engineers the cooperative *in vivo* context that expedites the folding process, ultimately yielding experimentally validated structures. The resulting dynamics provide mechanistic underpinnings to empirical models previously developed to fit kinetic experimental data.

A.1.2 Transformers Empowering Molecular Dynamics

A.1.2.1 Propagating the Topological Dynamics with Transformers

To reduce the informational burden in MD integration steps, we first simplify the dynamics taking into account the inherent topology of the MD-steering force field. The procedure known as quotient-space projection has been previously introduced [16–18] and is described in Chapter 5, so it will only be described cursorily for completion, emphasizing the adaptation required to implement an AI model of the *in vivo* molecular reality. To describe the projection onto quotient space, we consider conformations described by the backbone torsional coordinates $\{\phi_n, \psi_n\}_{n=1,...N}$ for a protein chain of length N. Conformation space becomes a $2N$-dimensional torus, i.e., the product of $2N$ unit circles, one for each dihedral coordinate of the protein backbone [9, 12, 16–18]. To coarse-grain atomistic MD runs, we introduce an equivalence relation "~," where two conformations are regarded as equivalent if and only if the torsional states of the individual residues lie within the same basins of attraction in the local potential energy surfaces, i.e., the Ramachandran plots. The Ramachandran basins are the allowed low-energy regions available to backbone torsional coordinates for each residue (Figure A.1) [16]. The rigorous mathematical construction of the modulo-basin metamodel simplification is given in the following section.

The "modulo-basin" coordinate representation constitutes a "quotient space" made up of equivalence classes. In generic terms, if x indicates the torsional state of the chain, the modulo-basin state to which x belongs is the class denoted $\overline{x}$, expressed as the Cartesian product of N Ramachandran basins designating the basin occupancies for the individual

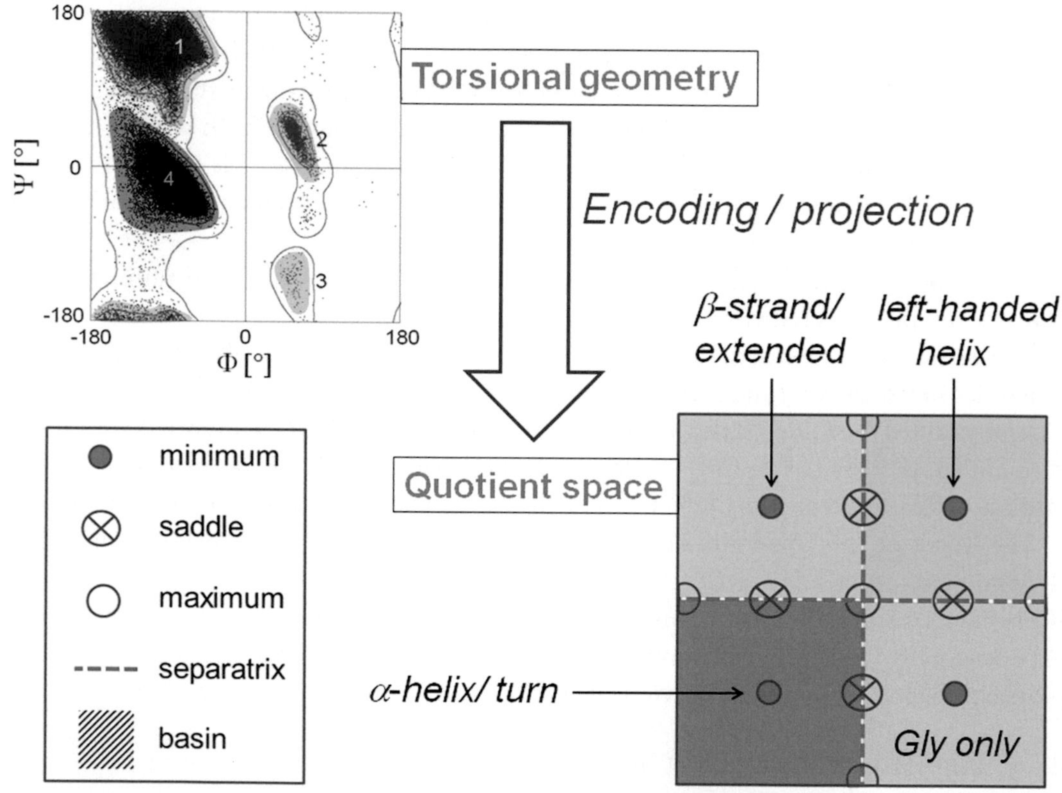

FIGURE A.1 Modulo-basin topological representation of the backbone torsional state of a protein chain. The topological representation of the Ramachandran energy plot steering the backbone torsional dynamics of a single residue is shown in the lower panel. The allowed regions in local torsional space are displayed in the upper panel. Opposite sides of the square are identified as per the ±180° identification of ϕ, ψ dihedrals, yielding a 2-torus, or Cartesian product of two circles. The four colored sectors morph topologically into the Ramachandran basins of potential energy subsuming the local conformational motifs, such as helix, turn, and β-strand. Except for glycine (G), the gray sector is energetically inaccessible to other residues due to steric clash with the side-chain. Other accessibility restrictions apply to geometrically constrained residues like proline (P) and its adjacent residues. Thus, the local 2-torus geometry is canonically projected onto the local topology with its organization of critical points.

residues: $\bar{x} = \prod_{n=1}^{N} b(x,n)$, where $b(x,n)$ is the basin of attraction occupied by residue n in conformation x. Thus, for any two conformations x, y: $x \sim y \Leftrightarrow \bar{x} = \bar{y}$, and the states in quotient space may be designated by encoding vectors. It is worth noting that the modulo-basin representation of the backbone torsional state of the chain is a "textual" representation, where the alphabet is comprised of four "letters" indicating the basin assignment at each position on the chain. This text can be processed by a transformer sequence-to-sequence (Seq2Seq) architecture [15], so the sequence transduction represents a broad conformational move arising once fast motions (intra-basin transitions) are averaged out.

We may project the MD trajectory on quotient space and then adopt transformer learning to propagate the textual modulo-basin metamodel of the dynamics to make computations more agile. This text entails a considerable simplification of conformation space, while the transition timescale for each iteration is the minimal Ramachandran intra-basin equilibration time τ=100ps, considerably larger than the femtosecond integration time step in MD. Since intra-basin events are averaged out in the modulo-basin representation and the iteration time step is five orders of magnitude longer, the propagated modulo-basin dynamics has a far better chance of capturing infrequent inter-basin transitions, a significant hurdle in MD modeling of *in vitro* protein folding.

The transformer text processing needed to propagate the modulo-basin dynamics uses a maximum likelihood *ansatz* to train the transformer so that the maximum likelihood weight, $f_\tau^{(t)}(n, b \rightarrow b')$, for basin transition $b \rightarrow b'$ at chain position n after time τ elapsed from time t, is influenced by the amino acid identity and basin occupancy at flanking residues ($n-2$, $n-1$, $n+1$, $n+2$), with an influence field in the LSTM (long short-term memory) progressively dilated, so that flanking positions ($n-3$, $n-2$, $n+2$, $n+3$), ($n-4$, $n-3$, $n+3$, $n+4$),..., ($n-8$, $n-7$, $n+7$, $n+8$) with respective attention gaps g=1,2,...,7 also impact the transition probability. Although both are grounded in the maximum likelihood estimation of basin transition probabilities using the same training data, this representation of the receptive field is different from a previous one that yields identical results in the *in vitro* context [17]. The previous "global" method learns to estimate transition probabilities between modulo-basin states of the whole chain, while the current approach determines basin transition probabilities at the residue level using semilocal influence fields. In this way, the number of neurons in the hidden layers is significantly reduced. This simplification is essential to reduce computational costs, as the transformer learns to incorporate the complex *in vivo* reality.

Thus the maximum likelihood weight of the basin transition becomes:

$$f_\tau^{(t)}(n, b \rightarrow b') = \prod_{g=1}^{L} w(n, g, B(t)) f(n, g, b \rightarrow b') \tag{A.1}$$

where $f(n, g, b \rightarrow b')$ gives the frequency of the $b \rightarrow b'$ basin transition at position n after time τ elapsed, obtained by projecting 100 "short" 220ns-MD runs onto a modulo-basin

τ-discretized time series for a specific chain composition giving the amino acid identity of residues $n-g-1$, $n-g$, n, $n+g$, $n+g+1$ in the chain, and basin assignment for residues $n-g-1$, $n-g$, n, $n+g$, $n+g+1$ determined by $B(t)$, the chain basin occupancy at time t. The different g-attention fields ($n-g-1$, $n-g$, n, $n+g$, $n+g+1$) bearing on the basin transition at position n are weighted in accord with attention weights $w(n,g,B(t))$, which are optimized vis-à-vis the loss function of the transformer.

A first stage in the transformer training yields transition frequencies $f(n,g,b \rightarrow b')$ following the maximum likelihood scheme. The propagation of the modulo-basin trajectory (time series) in the maximum likelihood scheme is determined by the $b \rightarrow b'$ transition probabilities for generic position n for the jump $t \rightarrow t+\tau$: $p_{ML}{}^{(t)}(n,b \rightarrow b') = \dfrac{f_\tau^{(t)}(n,b \rightarrow b')}{\sum_{b''} f_\tau^{(t)}(n,b \rightarrow b'')}$, where b'' is any of the four basins. The destiny basin b' after a time period τ is determined by a Monte Carlo scheme taking into account the four transition probabilities at each position n along the chain.

The input in the transformer is thus expanded through L hidden layers in accordance with attention range g for the g-quintuples centered at each position n on the chain and influencing the basin transition at position n. The transformer input is a sequence of basin assignments for each residue along the chain, and its output is the result of the sequence transduction resulting in the basin assignment after time τ has elapsed. To apply the attention operation, a modulo-basin state for a chain of length N may be labeled with a $4N$-sequence $(\mathbf{b}_1|\mathbf{b}_2|\ldots|\mathbf{b}_n\ldots)$ consisting of N binary 4-tuples $\mathbf{b}_n$ $(n=1,2,\ldots,N)$ indicating the Ramachandran basin occupancy for each residue, so the value 1 at entry m in $\mathbf{b}_n$ $(b_{nm}=1)$ indicates that residue n occupies basin m $(m=1,2,3,4)$ (Figure A.1). For example, (1000|0010) indicates the coarse-grained state of a dipeptide with the first residue in basin 1 and second residue in basin 3. Thus, the attention contribution has associated a convolutional kernel K(n, g, $B_K(n,g)$) that becomes a g-quintuple of binary 4-tuples (basin assignment $B_K(n,g)$ with range g, with the convolution operation carrying the Kronecker-delta factor $\delta_{A(n,g),\, A_K(n,g)}$, where $A_K(n,g)$ is the kernel amino acid assignment for positions $n-g-1$, $n-g$, n, $n+g$, $n+g+1$. Each sublayer $L(q,g)$ is labeled by the basin quintuple, in turn labeled by index $q=1,\ldots, 4^5=1024$, and attention gap g. In each sublayer, the convolution kernels are slid along the chain as stencils, and full coincidence at position n in $L(q,g)$ yields concatenation (enrichment) at position n in the (q,g)-feature map with the four attributes $f(n,g,b \rightarrow b')$ for the four *a priori* possible b' destiny basins.

After processing across all L hidden layers, we arrive at a full representation of the chain at layer L, where each position n is endowed with $L\times 4$ transition parameters $f(n,g,b \rightarrow b')$, one for each attention gap g and each destiny basin b'. In the $L+1$ layer, the products $\prod_{g=1}^{L} w(n,g\ ,B(t))\, f(n,g,b \rightarrow b')$ get computed, and the four transition probabilities $p_{ML}{}^{(t)}(n,b \rightarrow b')$ are obtained for each of the four *a priori* destinies b'. Finally, at the output, the destiny basin for each residue after time τ has elapsed from its original basin occupancy at time t is chosen from the Monte Carlo scheme.

A.1.2.2 Topological Metamodeling

In accordance with the notation and conceptual framework delineated in Chapter 5, the architecture of the underlying neural network that enables the topological metamodel discovery consists of two autoencoders, $AE1:(\gamma,\mu,M); AE2:(\pi,\mu^{\#})$, and the transformer that generates the map Γ (Figure A.2a). Thus, autoencoder AE1 discovers the latent manifold Ω of torsional dihedral coordinates representing the relevant internal degrees of freedom of the chain, whereas AE2 discovers the latent quotient manifold $\Omega/\sim$ of "modulo-basin" classes that provides the topological "textual" framework of the metamodel. The autoencoder AE2 is not optimized for the topological dynamics subsumed in the map $\Gamma: \Omega/\sim \rightarrow \Omega/\sim$. This map is obtained by training and optimizing the transformer described in the previous section. The variational optimization of the transformer leads to the fulfillment of the commutativity relations: $\mu^{\#} \circ \Gamma = F \circ \mu^{\#}, \pi \circ M = \Gamma \circ \pi$, that make it dynamically compatible with AE1 and AE2, as shown in Figure A.2a.

The task flow for the transformer network capable of computing the *in vivo* folding of a protein chain through text processing is described schematically in Figure A.2b.

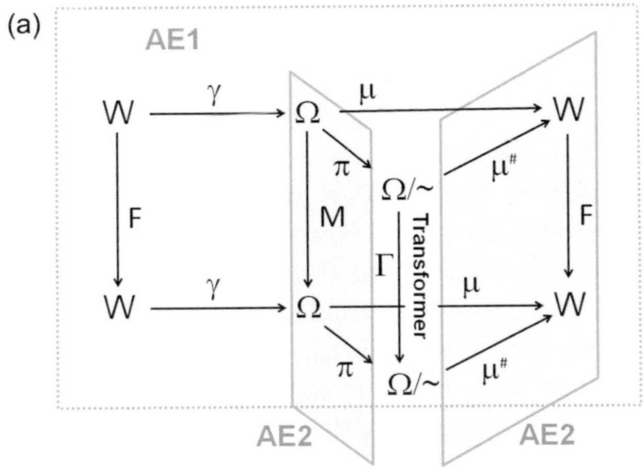

FIGURE A.2 Neural network architecture and task flow for textual processing/propagation of the topological dynamics metamodel of *in vivo* protein folding. (a) Neural network architecture for topological model discovery, consisting of two autoencoders, AE1 and AE2, and a transformer that generates the topological dynamics map Γ in the latent quotient manifold. (b) Task flow for a transformer capable of propagating short atomistic MD runs to generate the *in vivo* folding of a protein chain as "text processing." The "text" is the sequence indicating the assignment of a Ramachandran basin to each residue along the chain. The alphabet consists of 20x4=80 "letters," each specifying amino acid type and basin occupancy.

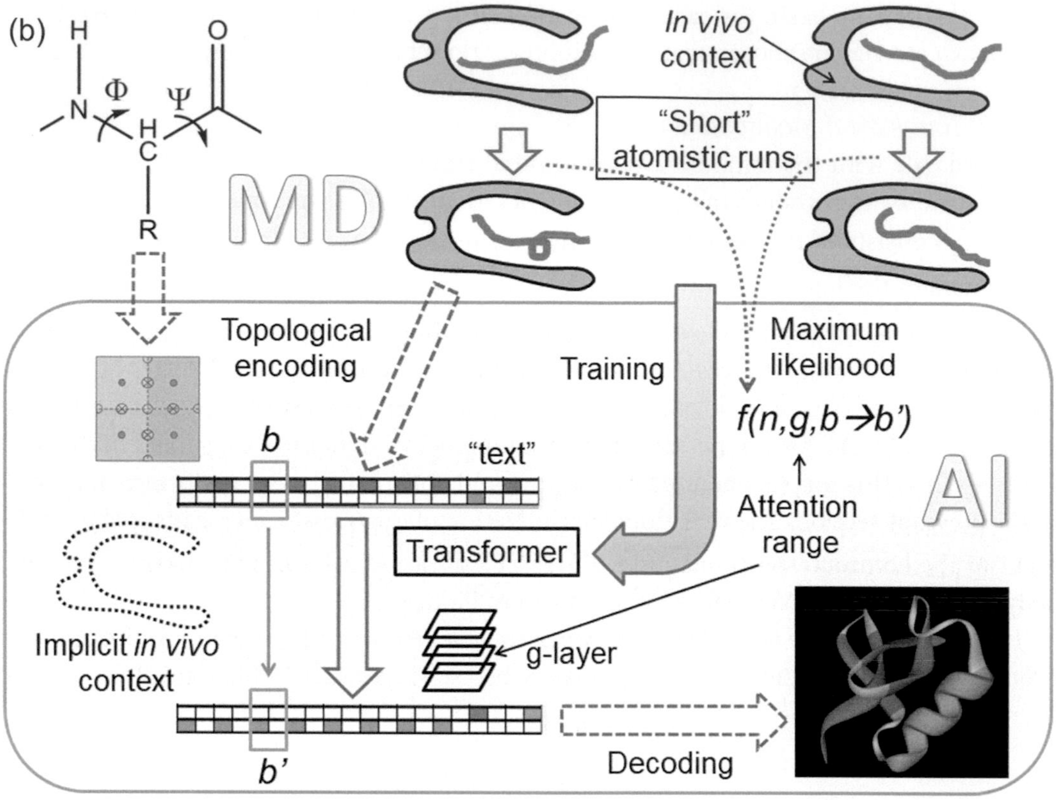

FIGURE A.2 Continued

The optimization of the attention-defining weights $w(n,g,B(t))$ is carried out during the second training phase for the modulo-basin propagator Γ_τ defined by the Monte Carlo generator of basin transitions. To train the transformer in this second stage, we break up the MD trajectories spanning a time period $[0, t_f]$ into two regions, a training portion with a timespan $[0, t_0]$ and an optimization portion covering the time interval $(t_0, t_f]$. With the trajectory statistics drawn for the region $[0, t_0]$, we obtain the basin transition frequencies for triads with different strides centered at each particular residue. On the other hand, a variational optimization process during the period $(t_0, t_f]$ enables us to determine the w-parameters. Thus, the *w*s are obtained from a commutativity condition that reflects the compatibility of fine and coarse-grained propagation of the trajectory and must be fulfilled during the period $(t_0, t_f]$. The commutativity condition is $\Gamma_\tau \circ \pi = \pi \circ M(\tau)$ as depicted by the following scheme:

$$\begin{array}{ccc} x(t) & \stackrel{\pi}{\rightarrow} & \overline{x(t)} \\ \downarrow M(\tau) & & \downarrow \Gamma_\tau \\ x(t+\tau) & \stackrel{\pi}{\rightarrow} & \overline{x(t+\tau)} \end{array}$$

Here π is the canonical projection – effectively, the encoding – that assigns the Ramachandran basins to each backbone torsional state of the chain (Figure A.1), and $M(\tau)$ is the molecular-dynamics propagator of torsional states of the chain. In other words, the parametrization of the modulo-basin propagator Γ_τ is optimized to minimize the loss function $\mathcal{L}(\Gamma_\tau)$ given by

$$\mathcal{L}(\Gamma_\tau) = M^{-1} \sum_{q=1}^{M} \| \pi x(t_0 + q\tau) - [\Gamma_\tau]^q \pi x(t_0) \|^2, \tag{A.2}$$

where $M\tau \leq t_f - t_0 < (M+1)\tau$. The choice of loss function is correct since $\mathcal{L}(\Gamma_\tau) = 0$ if and only if the propagator makes the diagram commutative. Once the optimal $\Gamma_\tau^* = \arg\min \mathcal{L}(\Gamma_\tau)$ has been obtained from stochastic gradient descent, the modulo-basin projected trajectory can be propagated beyond MD-accessible timescales. The set $\wp(t_0, t_f)$ of MD-generated states during the period $t_0 \leq t \leq t_f$ is sampled randomly at each iteration to compute the gradient with minibatches given by trajectory fragments spanned by randomly chosen time intervals of size $10^{-2}|t_f - t_0|$. To optimize the network connectivity that determines Γ_τ, the ADADELTA optimization protocol with a learning rate of 0.3 is adopted [19]. Physically meaningful timescales may be reached by computing the coarse states $\overline{x(t_f + q\tau)}$ beyond the simulation timespan t_f as $[\Gamma_\tau^*]^q \overline{x(t_f)}$.

A.1.3 Folding Proteins with Transformers

We now show that the protein folding dynamics may be reduced to an encodable dynamical system (EDS), characterized by an asymptotic behavior that is topologically described by a finite number of bits of information. To represent the protein chain dynamics as an EDS within a transformer platform (cf. Figure A.2), it is necessary to prove that we can provide a textual representation of the enslaving modes that describe the backbone conformation dynamics (Figure A.1) and subsume the determinant topological features of the force field [20]. An example of textual display of the backbone (Φ,Ψ)-representation would be a coarse-grained discretization that defines for each residue the local topology of the chain [9, 16].

For chain length N, conformation space Ω becomes the Cartesian product of N differentiable compact manifolds: $\Omega = \prod_{n=1}^{N} \Omega_n$, where Ω_n is a 2-torus spanned by the dihedral (Φ_n,Ψ_n)-coordinates of the n-th residue. Let $\mathcal{M} = \mathcal{M}(\{\Phi_n,\Psi_n\}_{n=1,\ldots,N}) = -\nabla U(\{\Phi_n,\Psi_n\}_{n=1,\ldots,N})$ denote the vector field defined on Ω and associated with the force-field potential energy U that governs the MD simulation of the folding process. The trajectories defined by the vector field $\mathcal{M}$ are τ-incrementally determined by $M(\tau)$, the propagator of torsional dynamics defined in Section A.2. Let $\pi_n\mathcal{M} = \mathcal{M}_n : \Omega_n \to T\Omega_n$ denote the pseudo-projection of $\mathcal{M}$ onto Ω_n, with $T\Omega_n$ denoting its tangent bundle. The map $\pi_n\mathcal{M} = \mathcal{M}_n$ assigns to each point $y=(\Phi_n,\Psi_n)$ in Ω_n the projections onto $T\Omega_n$ of all the vectors $\mathcal{M}(z)$ in $T\Omega$ associated with all points $z=(\{\Phi_{n'},\Psi_{n'}\}_{n'=1,\ldots,N})$ such that $(\Phi_{n'},\Psi_{n'})=(\Phi_n,\Psi_n)$ for $n'=n$, that is, all the points z in Ω that project onto y. In general, this pseudo-projection is a multi-valued map. To define a true projection, we define the blueprint of $\mathcal{M}_n$ on the local Ramachandran potential energy function $U_{R,n}$: $\Omega_n \to R$ (R=real numbers) as $\mathcal{M}_{R,n}(x) = \{\int[\delta(\alpha)\mathcal{M}_{n,\alpha}(x)]\, d\alpha\}\hat{e}(x)$, where $\hat{e}(x) = -\nabla_n U_{R,n}(x) / \| \nabla_n U_{R,n}(x) \|$ if $\nabla_n U_{R,n}(x) \neq 0$, $\hat{e}(x) = 0$, otherwise $\alpha = \arg\cos[\widehat{\mathcal{M}_{n,\alpha}(x).\hat{e}(x)}]$ for $\hat{e}(x) \neq 0$, and $\delta(\alpha)$ is the Dirac delta.

We now introduce the following.

Definition. A smooth (i.e., class C^1, with continuous first derivative) vector field V defined on a differentiable compact manifold is *textually encodable* (cf. Figure A.1) if it can be approximated arbitrarily closely by a C^1 vector field W topologically determined by the basins of attraction of a finite set $x_1, x_2, \ldots, x_J$ of generic singular points. A singular point x is generic if $W(x)=0$ and $Re\lambda(x) \neq 0$, for all λ=eigenvalue of the Jacobian matrix of W at x, and the basin of attraction of a singular point x_i is the set of points x whose destiny or omega set ($\omega(x)$) along the trajectory or integral line defined by W is x_i ($\omega_V(x)=x_i$). The distance d(V,W) between two vector fields V and W is defined via $d(V,W) = \int \| V(x) - W(x) \|^2 \, dx$ where integration extends over the domain manifold common to both fields.

According to the definition of textually encodable, the flow defined by V can be approximated arbitrarily closely by a flow W coarsely defined by transitions between basins of attraction of J singular generic points. This implies that the flow V becomes encodable as text, that is, the generic point x may be represented by a binary vector $\bar{x}$ that simply specifies the basin of attraction relative to W that contains x.

To enable transformer technology to extend the molecular dynamics of unassisted and chaperone-assisted protein folding (Figure A.2), we need to prove the following.

> *Theorem A.1.* The Ramachandran vector fields $\mathcal{M}_{R,n}(x)$ (n=**1**,…,N) are textually encodable, and hence the propagator (flow) $M(\tau) = \exp(\tau\mathcal{M})$ represents an EDS.

Proof. Since Ω_n is compact (a 2-torus), we simply need to prove that $\mathcal{M}_{R,n}$ can be approximated arbitrarily closely by a vector field Y with generic singularities (saddles, sinks, and sources). If they are generic, they are isolated by definition, and hence, since they do not accumulate, they must be finite in number. Since the singularities are finite, the vector field $\mathcal{M}_{R,n}$ becomes textually encodable, since the coarse-graining of Y only requires that we provide the spatial organization of the generic singularities of Y and the partition of Ω_n into a finite number of basins of attraction of sinks (two-dimensional) and saddles (one-dimensional, known as separatrices). An illustration of a textual encoding of the Ramachandran map is given in Figure A.1.

Essentially, we need to prove that for a given arbitrarily small ε>0, we can find a vector field Y with a finite number of generic singularities satisfying: $\int \| \mathcal{M}_{R,n}(x) - Y(x) \|^2 \, dx < \varepsilon$, where integration extends over Ω_n.

The vector field $\mathcal{M}_{R,n} : \Omega_n \to T\Omega_n$ is a differentiable cross-section of the tangent bundle $T\Omega_n$. The mapping $\mathcal{M}_{R,n}$ can be approximated by a map L: $\Omega_n \to T\Omega_n$, with L satisfying two conditions [21]: (a) $\int \| \mathcal{M}_{R,n}(x) - L(x) \|^2 \, dx < \varepsilon / 2$; and (b) L is transversal to Ω_n, meaning that at each singular point of L, the Jacobian matrix is non-singular ($\lambda \neq 0$), or the singularities of L are simple. The existence of L is guaranteed by Thom's transversality theorem [21], which posits that transversal maps are dense in the space of differentiable vector fields on a compact manifold and the observation that if a map contains simple singularities, there must be a finite number of them because (a) Ω_n is compact and (b) simple singularities are by definition isolated. By means of a small C^1-perturbation with norm $<\varepsilon/2$, we can turn L into the vector field Y with the desired following properties:

a) Y has a finite number of singularities, all generic, and

b) $\int \| \mathcal{M}_{R,n}(x) - Y(x) \|^2 \, dx \leq \int \| \mathcal{M}_{R,n}(x) - L(x) \|^2 \, dx + \int \| L(x) - Y(x) \|^2 \, dx < \varepsilon$

As shown subsequently, this approximation theorem ensures the applicability of the transformer platform to extend molecular dynamics spanning realistic timescales in accord with the scheme given in Figures A.1 and A.2.

We have shown that for any n, the Ramachandran map $\mathcal{M}_{R,n}$ is the accumulation point of a sequence $\{Y_j\}_{j=1,...}$ of "simple" vector fields with a finite number of generic singularities on Ω_n. The modulo-basin quotient space $\Omega_n/[\sim]_Y = \Omega_n / Y$ with respect to vector field Y is defined by the map $x \rightarrow \omega_Y(x)$ [22, 23]. As we approximate $\mathcal{M}_{R,n}$, there is a j-value $j{=}j^*$ beyond which the quotient space remains essentially invariant (if it would change, we would deviate from approaching $\mathcal{M}_{R,n}$). In other words: $\forall n \exists j^*(n): \Omega_n / Y_j \approx \Omega_n / Y_{j'} \, \forall j > j' > j^*$, where $\approx$ denotes "isomorphic." Then, since for j large enough, each approximant vector field contains the same number K of generic singular points, we may textually encode the local state x of the chain by specifying the singular point x_k (k=1,...,K) that constitutes the omega set (point) of x in accord with the surjective map $x \rightarrow \omega_{Y_j}(x) = x_k \, \forall j > j^*$. Thus, a binary K-tuple with as many entries as generic singular points for $Y_j (j > j^*)$ serves as the textual encoding for the coarse-grained state $\bar{x}$, with 1 for the k-th entry of the encoding vector and 0 for all other entries.

A.1.4 Artificial Intelligence Recreates *In Vivo* Reality

We now incorporate the role of *in vivo* settings in steering the protein chain along an expeditious folding pathway. Given computational limitations, atomistic detail on a folding trajectory in an *in vivo* context, say, in the chaperone chamber, is hardly feasible. For a typical single-domain protein (N>50), such computations are not likely to reveal the mechanism by which the cellular context is able to expedite folding. To generate coarse-grained *in vivo* folding trajectories, we implement a deep learning system that captures the expediency of the *in vivo* environment when trained with accessibly short (30ns) MD simulations starting at unfolded or metastable states generated *in vitro*. We need to capture the means by which the *in vivo* setting removes kinetic traps by selectively disrupting misfolded states that would otherwise be susceptible to aggregation.

Let $\widetilde{\Gamma_\tau}$ denote the modulo-basin propagator that commutes with $\widetilde{M(\tau)}$, the MD operator that steers the chain torsional dynamics in the *in vivo* setting (in our illustration, the GroEL cavity). The propagator $\widetilde{\Gamma_\tau}$ is obtained by training the transformer with many short *in vivo* runs, and its domain $\mathcal{D}(\widetilde{\Gamma}_\tau)$ consists of basin assignments along the chain that undergo basin transitions (including retentions) when the chain explores conformations in the *in vivo* setting. The domain for Γ_τ^* is similarly defined *mutatis mutandis* in association with the *in vitro* setting.

To generate *in vivo* folding trajectories that span realistic timescales, a significant coverage is required of the influencing basin occupancies at positions flanking each residue along the chain placed in the *in vivo* environment. To that effect, we construct the propagator $\widehat{\Gamma}_\tau$ obtained by overwriting basin transitions yielded by Γ_τ^* in the domain $\mathcal{D}(\Gamma_\tau^*) \cap \mathcal{D}(\widetilde{\Gamma}_\tau)$, replacing them with the transitions dictated by $\widetilde{\Gamma}_\tau$. Thus, we get $\widehat{\Gamma}_\tau = \chi_{\mathcal{D}(\widetilde{\Gamma}_\tau)} \widetilde{\Gamma}_\tau \otimes \chi_{\mathcal{D}(\Gamma_\tau^*)\setminus[\mathcal{D}(\Gamma_\tau^*) \cap \mathcal{D}(\widetilde{\Gamma}_\tau)]} \Gamma_\tau^*$, where χ denotes the characteristic function (1 on its support, 0 elsewhere), and $\mathcal{D}(\Gamma_\tau^*) \setminus [\mathcal{D}(\Gamma_\tau^*) \cap \mathcal{D}(\widetilde{\Gamma}_\tau)]$ denotes the complement of $\mathcal{D}(\Gamma_\tau^*) \cap \mathcal{D}(\widetilde{\Gamma}_\tau)$ in $\mathcal{D}(\Gamma_\tau^*)$. Thus, the *in vivo* propagator $\widehat{\Gamma}_\tau$ commutes with $\widetilde{M(\tau)}$ in $\pi^{-1}[\mathcal{D}(\widetilde{\Gamma}_\tau)]$ and with $M(\tau)$ in $\pi^{-1}\ \{\mathcal{D}(\Gamma_\tau^*) \setminus [\mathcal{D}(\Gamma_\tau^*) \cap \mathcal{D}(\widetilde{\Gamma}_\tau)]\}$, implying that $\widetilde{\Gamma}_\tau$ overwrites Γ_τ^* in $\mathcal{D}(\Gamma_\tau^*) \cap \mathcal{D}(\widetilde{\Gamma}_\tau)$.

A.1.5 Propagating *In Vitro* Dynamics with Autoencoders

The efficacy of the quotient-space simplification is illustrated by computing folding pathways converging to native structures. We selected an N=57 chain known to fold autonomously in an *in vitro* setting: The thermophilic variant of the B1 domain of protein G from *Streptococcus* (PDB.1GB4). The thermophile was chosen over the wild type due to the higher stability of the folded structure. Using the CHARMM package (the free version of CHARMM) [3], we first generated a 220μs-folding trajectory within the NPT (isothermal/isobaric, T=298K) ensemble [16]. The system was equilibrated at 300K and 1 atm, and the runs were performed in the NPT ensemble with a Nosé–Hoover thermostat [24–26] and an MTK barostat [27], with the mass of all hydrogen atoms set at 4 amu, while the time step was fixed at 3.0fs. The full MD trajectory encoded as the modulo-basin version is shown in Figure A.3a, with selected t_0=120μs as the training parameter and $t_0 < t \leq 220\mu s$ as the learning period to optimize the propagator Γ_τ. The coarse-grained trajectory is then propagated up to t=7500μs (Figure A.3).

We note that the final state is stable, prevailing since the time of its inception at t=3108μs discerned in Figure A.3b. Furthermore, the decoded final stable state at t=7500μs (Figures A.3b and A.4), corresponding to the eigenvector associated with eigenvalue 1 of the matrix Γ_τ^*, is topologically equivalent to the native crystallographic state (PDB.1GB4), revealing the same pattern of antiparallel and long-range parallel β-sheets with α-helix packed against the β-sheet motif. The backbone-atom RMSD is estimated at 1.88Å. The decoding of modulo-basin states into torsionally specified conformations is obtained using a deep learning system previously described [16–18].

In the *in vitro* folding context, at least two metastable states can be spotted to be distinctively generated at 0.7 and 2.7ms (Figure A.3), corresponding to the chain conformations displayed in Figures A.4b and A.4c, respectively.

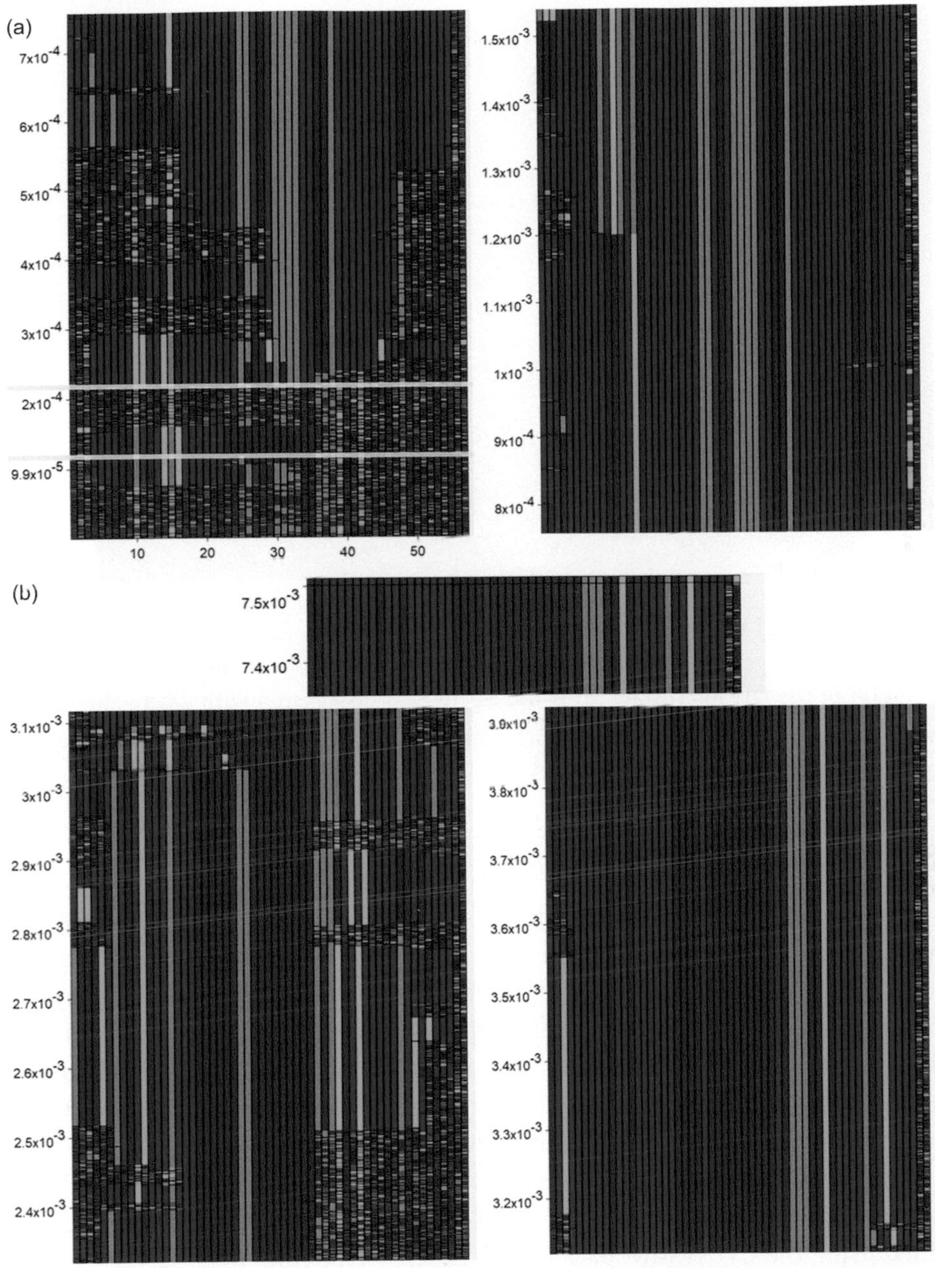

FIGURE A.3 Coarse-grained propagation generated by a DL system of an atomistic MD trajectory covering 220 microseconds. Key portions of the MD trajectory at 1μs-resolution covering timespan [0,7500μs]. The first 220 microseconds correspond to a coarse-grained representation of an MD trajectory at 1μs-resolution for the N=57 chain of the B1 domain of protein G. The trajectory is subsequently extended in time within a deep learning platform. At each time (seconds, vertical axis), the modulo-basin state of the chain is shown as a color sequence, where the basin assigned to each residue on the horizontal axis is specified in accordance with the convention set forth in Figure A.1. The MD trajectory was generated within an NPT (isothermal/isobaric, T = 298 K) ensemble, with interval [0, t_0=120μs] as the training region and interval [120μs, 220μs] as the learning period. On panel **b**, the steady state ensemble is shown for the final portion of the trajectory. The native-like steady state develops at around 3100μs.

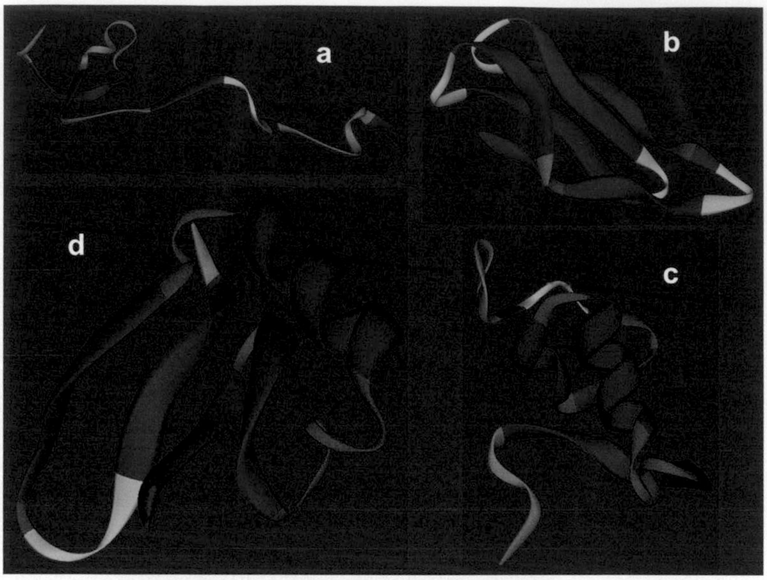

FIGURE A.4 Decoded torsional states in ribbon representation for the AI-enabled propagation of the MD trajectory, generated at start (a), 0.7ms (b), 2.7ms (c), and 7.5ms (d). The modulo-basin representation of the last microstate (d) is the eigenvector associated with eigenvalue 1 of matrix Γ_τ^* and is topologically equivalent to the native state from PDB entry 1GB4 obtained by X-ray diffraction crystallography, with RMSD=1.88Å.

A.1.6 Short Atomistic Simulation of *In Vivo* Protein Folding

This section is purely methodological and describes a standard setting for molecular dynamics in a highly complex molecular reality. A conformation identified with the X-ray diffraction structure of GroEL$(ATP)_{14}$ (PDB.1SX3) was adopted to model the GroEL chamber assumed to expedite the folding of the B1 domain of G protein (PDB.1GB4). Excess Mg^{++}-complexed ATP molecules were removed, while the 22 inherently disordered C-terminal amino acids, mainly comprised of Gly-Gly-Met repeats, were incorporated, as previously specified [24]. One protein chain was placed at the center of the GroEL cavity, and the system was solvated in a 200 Å × 200 Å × 200 Å water box at 120 mM KCl concentration. The final system was composed of ~1,108,000 particles. Simulations of 30ns were performed using the CHARMM package [3]. Seventy-two runs were performed with the substrate protein initially in the apo GroEL chamber with the second chamber in $(ATP)_7$-state. The substrate in its resultant conformation was subsequently transferred as a rigid body to the center of the $(ATP)_7$-chamber with the second chamber in $(ADP)_7$-state. Ten runs were started in an unfolded state, two in the protein native state (PDB.1GB4), and sixty runs were started in metastable states identified in *in vitro* runs, as illustrated below. The system was subject to energy minimization and equilibrated at 300 K and 1 atm with harmonic restraints on the alpha carbons. The runs were performed in the NPT ensemble with a Nosé–Hoover thermostat [25, 26] and an MTK barostat [27], with the mass of all hydrogen atoms set at 4 amu. The time step was fixed at 3.0 fs.

A.1.7 *In Vivo* Folding Trajectories Generated with Transformer Technology

To incorporate the influence of the *in vivo* reality, the transformer choreographs folding disruptions at states where $\widetilde{\Gamma}_\tau$ overwrites Γ_τ^* when the application of the latter generates metastable states characterized by the persistence of specific basin assignments. The transformer operationally incorporates the *in vivo* intervention step. Assume a substantial portion (typically 80% or more) of the basin assignment for the chain is detected to be locked at time $t_f + q_s\tau$, then the potentially disruptive propagator $\widetilde{\Gamma}_\tau$ overwrites Γ_τ^* for the step $t_f + q_s\tau \to t_f + (q_s + 1)\tau$. Thus the coarse state at $t_f + (q_s + 1)\tau$ is computed as $\widetilde{\Gamma}_\tau[\Gamma_\tau^*]^{q_s} x(t_f)$. The disruptive propagator excludes the basin retention $b'=b$ frequency $f_\tau(n,b,b)$ from the Monte Carlo scheme that assigns at time $t_f + (q_s + 1)\tau$ the destiny basin b' at chain position n. For residues that did not retain their basin assignment at time $t_f + q_s\tau$, the propagation $b \to b'$ is constructed as in Γ_τ^*.

A.1.8 AI Generates *In Vivo* Folding Pathways

To illustrate the generation of *in vivo* folding pathways, the AI platform constructed an *in vivo* reality by learning to generate the propagator $\widetilde{\Gamma}_\tau$ from 72 atomistic 30ns-MD runs of the N=57 protein chain exploring conformation space within the apo and $(ATP)_7$ states of the GroEL chamber. The full catalytic round in the apo GroEL chamber is allowed to be completed before the chain in its final conformation is placed in the $(ATP)_7$ chamber, as indicated in Figure A.5a–c,f,g. The transformer generated folding trajectory is shown in Figure A.5a, while Figures A.5b,c show the influence of the *in vivo* context depicted as selective interferences with basin transitions. The basin assignments implicating the *in vitro* model valid in $\mathcal{D}(\Gamma_\tau^*)\setminus[\mathcal{D}(\Gamma_\tau^*)\cap\mathcal{D}(\widetilde{\Gamma}_\tau)]$ are marked in white, those involving *in vivo* intervention in $\mathcal{D}(\widetilde{\Gamma}_\tau)\setminus[\mathcal{D}(\Gamma_\tau^*)\cap\mathcal{D}(\widetilde{\Gamma}_\tau)]$ are marked in pink, while black denotes *in vivo* overwriting of *in vitro* assignments in $\mathcal{D}(\Gamma_\tau^*)\cap\mathcal{D}(\widetilde{\Gamma}_\tau)$ and gray indicates coincidence between *in vitro* and *in vivo* assignments in $\mathcal{D}(\Gamma_\tau^*)\cap\mathcal{D}(\widetilde{\Gamma}_\tau)$. The starting conformation is shown in the ribbon rendering in Figure A.5d. The convergence to a native fold (Figure A.5e), with RMSD=1.91Å relative to PDB.1GB4, occurs 75 times faster than in the *in vitro* setting (cf. Figure A.3), attesting to the efficiency of the *in vivo* context to extricate the chain from kinetic traps during the annealing phase taking place in the apo state of the GroEL chamber.

Strikingly, ultra-expeditious *in vivo* folding, 750 times faster than *in vitro* folding, is generated by AI with a transformer further enriched with *in vivo* reality (Figure A.5f,g). This reality is extracted from 120 coarse runs spanning 100μs each generated by the transformer trained as specified above with the MD runs subsuming the GroEL context.

A.1.9 Reverse-Engineering of the Expeditious *In Vivo* Context I: Iterative Annealing in the Chaperone Chamber

The folding-assisting dynamics of the apo state GroEL chamber consist basically of a tightly choreographed stochastic annealing process, which is very different from the folding assistance provided by the $(ATP)_7$ state of the GroEL chamber. As the substrate is placed in the cage baricenter, the seven GroEL subunits are subject to conformational selection

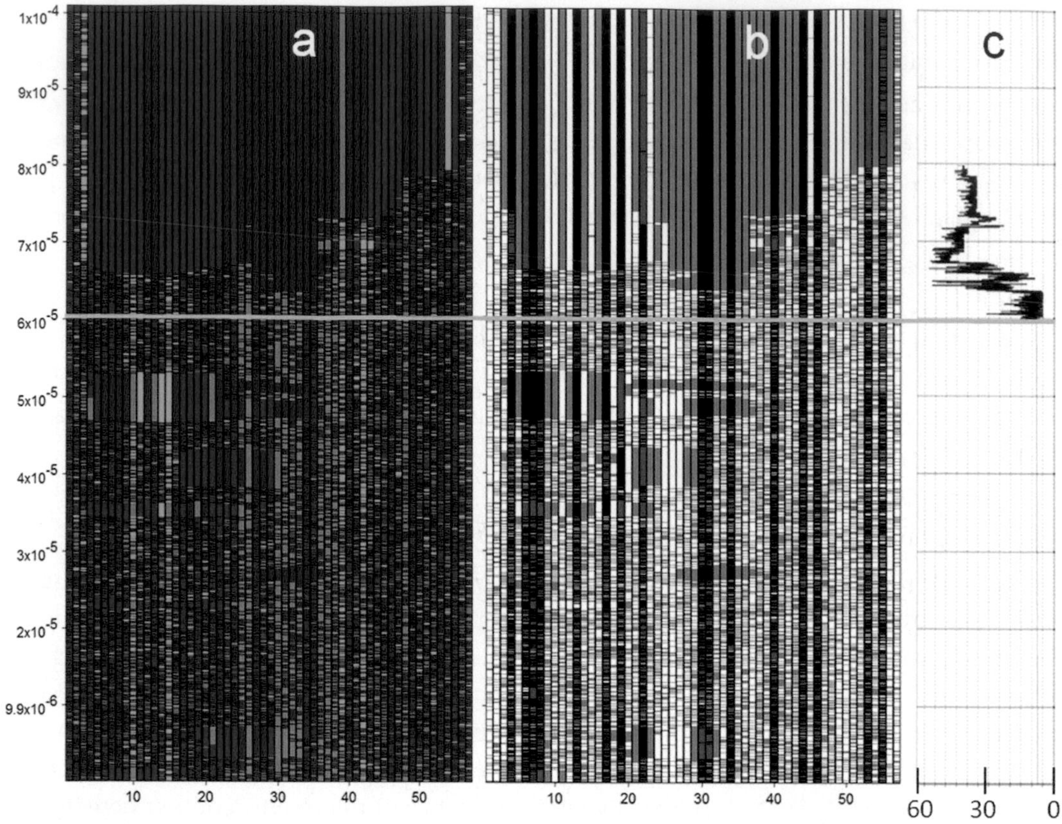

FIGURE A.5 Expeditious *in vivo* folding trajectories generated in an AI transformer platform (Figure A.2). The full catalytic round in the apo GroEL chamber is allowed to be completed before transferring the chain to the $(ATP)_7$ chamber. The time of transference is marked by the yellow line in Figures A.5a–c, f–g. (a) Coarse-grained trajectory at 0.1μs resolution generated with a CNN trained with 72 atomistic 30 ns-MD runs of the chain dynamics within a GroEL environment. (b) Influence of *in vivo* context, as described in the main text. (c) Burst of cooperative interventions of the *in vivo* context. The number of protein BHBs that are partially shielded from disruptive hydration by side-chain nonpolar groups in the flexible tails of the chaperonin, supplementing intramolecular BHB wrapping. (d) Initial state. (e) Decoded final state. (f) Ultra-expeditious *in vivo* folding trajectory at 0.1μs resolution generated by "wiser AI" with a CNN further enriched with *in vivo* reality from 120 runs spanning 100μs each and generated by the CNN trained as specified in (a). (g) Influence of *in vivo* context steering the ultra-expeditious folding pathway.

in both cases. Initially, in the apo state, the T-conformation of the subunit, reported in PDB.1XCK, prevails for all seven subunits and constitutes a grabber of the folding substrate, as the C-terminus flexible hydrophobic tails comprising the last 22 residues of the subunit chain (526-548, absent in PDB structure) are initially free to interact intramolecularly. The T-centered ensemble occurs right after ADP is hydrolytically removed from the equatorial domains, a chemical event not explicitly modeled by the transformer. The T-centered ensemble with α-carbon RMSD dispersion at 2.2Å for the initial 525 ordered residues prevails for ca. 4μs and is denoted T_{grab} (Figure A.6). However, after the substrate

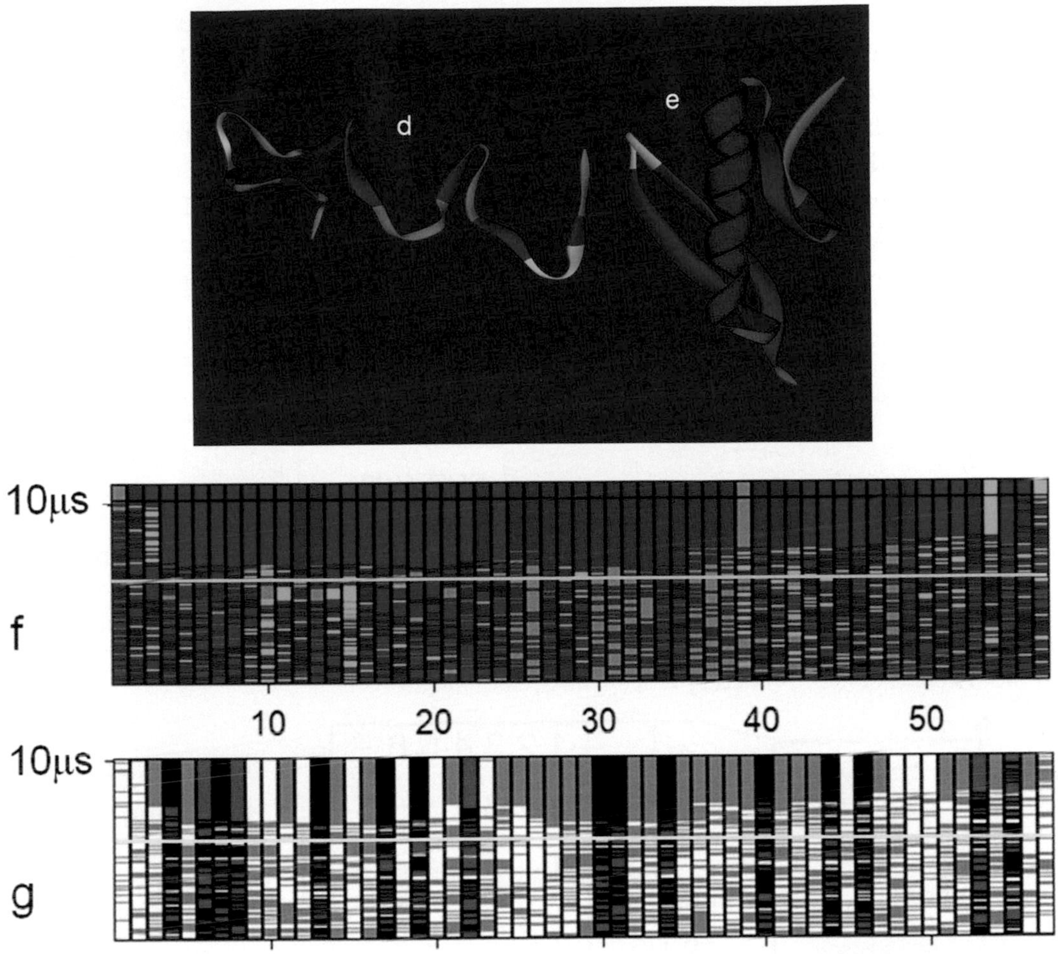

FIGURE A.5 Continued

is tightly held in the chamber, two conformational ensembles named Grab and Rel (short for "release") become selected for the seven subunits. The Grab and Rel ensembles are centered respectively at PDB structures 4KI8 ("R_{ADP}" state) and 1XCK (taut "T" state) with ensemble dispersions at 2.9Å and 1.8Å α-carbon RMSD, respectively. The substrate-grabbing role initially associated with the T_{grab} ensemble is now switched over to the R_{ADP}-centered ensemble Grab (Figure A.6). Again, the inherently disordered and hydrophobic 526-548 tails containing the GGM repeats and tethered to the C-termini of the equatorial domains are excluded from the RMSD calculations. In Grab, the C-terminus tails for subunits around the PDB.4KI8 conformation are free to interact with the substrate protein. By contrast, in Rel, the more rigid apical and equatorial domains interact intramolecularly through the hydrophobic C-terminus tails (Figure A.6), which are therefore unable to bind the folding substrate and preclude capping of the chamber at the apical region by the GroES unit [28]. The two ensembles are allosterically anticorrelated (Figure A.6a), so if one subunit is in Grab, the nearby unit transitions to Rel in a ~9.9μs timescale and vice-versa, as shown in Figure A.6a. Because the number of anticorrelated subunits is odd (7), there is always conformational frustration (Figure A.6b): One subunit tends to be in Grab and Rel simultaneously as it gets conflictive signals from its flanking subunits. This frustrated subunit slides through the ring (Figure A.6b), with the mismatched structure with RMSD at roughly 1/2 the RMSD between Grab and Rel (Figure A.6a). Even at an ensemble average level, the dynamics in the apo GroEL chamber reflects a breaking of the seven-fold symmetry at all times due to the frustration. The apo form "mechanically" dismantles misfolded kinetic traps, an assertion validated by contrasting Figures A.3 and A.5. This sort of

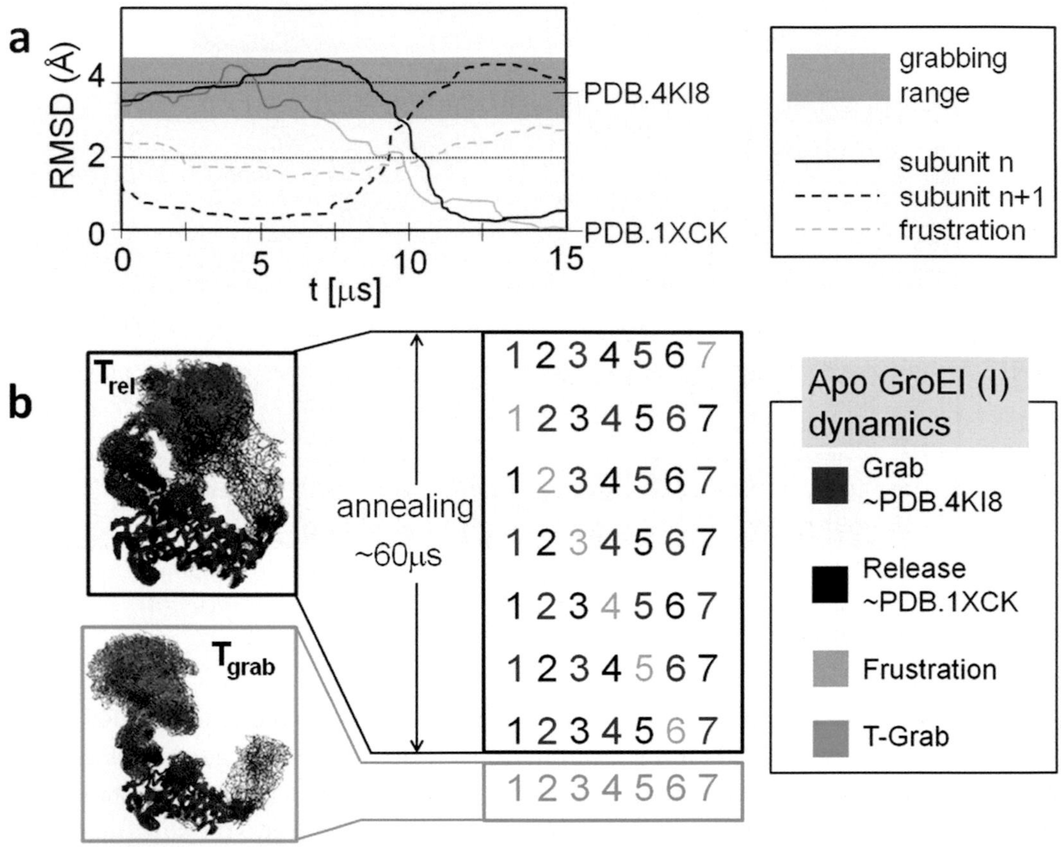

FIGURE A.6 Dynamics of a catalytic folding-assistance cycle of the GroEL chamber in the apo state with subunit conformational selection resulting upon binding to the folding substrate (cf. Figure A.5). Subunits are labeled by cyclic numerical indices (1, 2, …, 7, 8=1, 9=2, …). (a) Conformation dynamics reflecting allosteric anticorrelation across adjacent subunits (dark lines) and a pair including the frustrated subunit (gray lines). The RMSD is taken relative to PDB.1XCK. (b) The choreography of one complete annealing cycle determined by the alternating subunit roles of grabbing and releasing, and the progressive displacement of the conformationally frustrated subunit along the annulus. The Grab (Rel) ensemble adopted by a subunit promotes transition to the Rel (Grab) ensemble in the adjacent subunit. Frustration is inevitable due to the odd number of subunits and their annular assemblage.

intermittent annealing has been previously postulated in prescient kinetic models that fit experimental data [28] and now finds support at the molecular-dynamics level.

The annealing interactions between the 526–534 C-terminus tails and the folding substrate are stochastic in nature and cyclically choreographed with alternating subunit participation to disrupt metastable states. Examples of such metastable states are shown in Figures A.4b,c. They constitute kinetic traps if the folding process were to take place in the test tube. With this analysis, the molecular underpinnings and mechanistic realization of the prescient "iterative annealing model" [28] are brought to light.

A.1.10 Reverse-Engineering of the *In Vivo* Context II: The GroEL Chamber in the $(ATP)_7$ State

By contrast with the GroEL chamber in the apo state, the $(ATP)_7$ chamber has only one conformational ensemble selected for its seven subunits upon the incorporation of the folding substrate. This is a far more flexible and diverse ensemble than those observed for the apo chamber and is centered at PDB structure 1SVT (R-state), with maximum RMSD within the ensemble at 4Å for α-carbon resolution. This conformational selection is consistent with the previously established fact that the T→R transition is triggered by ATP binding to the equatorial domain [28]. Using the flexible hydrophobic tails tethered at subunit C-termini, the $(ATP)_7$ chamber scaffolds folding-nucleating conformations by wrapping native-like backbone hydrogen bonds (Figure A.5c), preventing structure disruption through backbone hydration.

Reverse-engineering the *in vivo* folding context is tantamount to elucidating how the *in vivo* environment cooperates with the folding process, making it expeditious. To accomplish this goal, we examined how a budding native-like secondary structure that would not prevail *in vitro* becomes protected from disruptive backbone hydration in the $(ATP)_7$ chamber, enabling it to nucleate the formation of the native structure (Figure A.5c). Since the main determinant of secondary structure is the backbone hydrogen bond (BHB), we decoded the modulo-basin representation into a tensor of three-body correlations (i,j,k), whereby a third residue (k) becomes wrapper of the BHB pairing residues i and j [16, 18, 25]. By "wrapper," we mean capable of excluding water molecules that may otherwise form hydrogen bonds with BHB-paired residues i and j, thereby locally disrupting the protein structure. In practice, a residue k becomes an (i,j)-wrapper when it contributes side-chain nonpolar groups (CH_n, n = 1,2,3) located within a BHB microenvironment consisting of two intersecting balls of radius 6Å centered at the α-carbons of the paired residues i, j [16, 18, 29, 30].

For proteins with reported structure, the extent of BHB wrapping, w, is identified from structural coordinates within the PyMol platform using a PyMol plugin [30]. The local parameter w gives the number of side-chain nonpolar groups contained within a predetermined BHB environment. A batch-mode wrapping analysis of the PDB (download February 7, 2022) revealed that 86% of BHBs satisfy $w=26.6\pm7.5$, while all reported BHBs satisfy $w\geq11.6$. The last relation indicates that the sustainability of a BHB requires wrapping values higher than two standard deviations below the mean.

To reverse-engineer the *in vivo* context, we decode the modulo-basin state $\overline{x(t)}$ into a wrapping (*ijk*)-tensor $\boldsymbol{L} = \boldsymbol{L}(x(t))$ through adaptive learning. To that effect, we use a dilated CNN with weighted connectivity vector θ that infers BHBs and their respective wrapping in the feature-extraction stage [16, 29]. Feature extraction operates by generating torsional conformations compatible with $\overline{x(t)}$ in layers filtered with receptive fields scaled by dilation parameters that correspond to contour distances ($|k-i|$, $|k-j|$) between wrapping residue (k) and BHB-paired residues (i,j).

As the flow progresses toward CNN output, the feature maps from hidden layers reveal longer and longer ranges of wrapping correlation. The decoding CNN adopts the loss function

$$J(\theta) = M^{-1}\sum_{q=1}^{M} \| \boldsymbol{L}\left(x\left(t_0 + q\tau\right)\right) - \boldsymbol{L}_\theta\left(\left[\widehat{\Gamma_\tau}\right]^q \overline{x\left(t_0\right)}\right) \|^2,$$

where $M\tau \leq t_f - t_0 < (M+1)\tau$, $\|.\|$ is the Frobenius norm, and $\boldsymbol{L}_\theta(\overline{x(t)})$ is the inferred (learned) wrapping tensor associated with the modulo-basin coarse state $\overline{x(t)}$. Thus, the minimization of $J(\theta)$ is a least-squares problem numerically solved with stochastic gradient descent. To train the network, the ADADELTA method with a learning rate of 0.3 is adopted [19, 29].

Strikingly, during the nucleation period $60\mu s \leq t \leq 80\mu s$ (Figure A.5a), *none* of the BHBs in the wrapping tensor $\mathbf{L}_\theta(\overline{\mathrm{x(t)}})$ inferred by the optimized transformer satisfies $w \geq 11.6$, making the nucleus thoroughly exposed to disruptive backbone hydration and therefore unstable unless exogenous BHB wrapping takes place. That means that the sustainability of the nucleating BHBs results from a burst of intermolecular protection of the protein conformations by the chaperonin during the structure-nucleating period $60\mu s \leq t \leq 80\mu s$ (Figure A.5c). This burst implies extensive exogenous contribution to the wrapping of the BHBs that eventually steer the formation of the native structure at $t > 80\mu s$. This finding suggests a nucleation mechanism that is unattainable in an *in vitro* setting due to insufficient wrapping or overexposure of the nucleus to disruptive hydration. *The GroEL chamber in the (ATP)$_7$ state intervenes cooperatively and stochastically to stabilize the folding nucleus, thereby committing the chain to fold.*

A.1.11 Can AI Truly Handle *In Vivo* Molecular Reality?

AI is revolutionizing molecular biophysics at a fast pace. Using AI, it has become possible to predict protein structures of amino acid sequences with staggering accuracy [1], the holy grail in the field a few years ago. One of the next frontiers involves predicting folding pathways and assessing the role of *in vivo* molecular contexts in expediting the folding process. This appendix represents a first step in that direction and shows that AI may fulfill the expectations through the construction of a metamodel. The effort is justified because natural proteins have been evolutionarily selected to fold in a cellular environment, not *in vitro*. It becomes imperative to learn how the cellular setting assists the folding process, prevents aberrant aggregation resulting from high local concentrations, and disrupts metastable states that may constitute kinetic traps. This effort may broaden the technological base of the pharmacological industry since drug targeting efficacy ultimately needs to be assessed *in vivo*, not in the test tube, and drug-induced folding requires an assessment of time-dependent molecular processes, not merely of static assemblages [18, 29].

Computational efforts to generate *in vitro* folding trajectories are commendable and provide useful insights [8, 10, 11] whenever an Anfinsen scenario [7] can be upheld but may arguably be misplaced in view of the fact that natural proteins have evolved to fold in an *in vivo* context. On the other hand, given the wanton complexity of cellular settings, the generation of *in vivo* folding pathways becomes unfeasible with current computational technologies. To address these challenges, this appendix introduced a transformer metamodel to recreate an *in vivo* reality with the goal of reverse-engineering the molecular underpinnings of folding expediency. We built a transformer NN trained with numerous short (30ns) MD runs to incrementally incorporate the very *in vivo* complexity that precluded MD from accessing realistic timescales and from capturing rare folding events. Atomistic MD simulations are unlikely to generate realistic *in vivo* folding pathways, even on dedicated supercomputers, but they can train an AI system such as the one presented in this appendix to enable the reverse-engineering of cooperative *in vivo* contexts that expedite the folding process.

As an illustration, the transformer technology is deployed to reverse-engineer the steering participation of the ATP-consuming chaperone GroEL in the protein folding process [4–6]. We observe that the apo and $(ATP)_7$ states of the GroEL chamber play different roles as they contribute to expediting the folding process. The first acts stochastically to disrupt budding misfolds, with allosteric anticorrelated structural transitions between adjacent subunits taking place while a structural defect arising from conflictive flanking signals travels along the annulus and completes its cycle in about 60μs. Thus, in the apo state, the GroEL chamber is shown to mechanically disrupt misfolded states in a cyclic choreography of alternating participating subunits. This choreography of alternating participants provides, at least to a degree, the molecular underpinnings of the prescient empirical model of "iterative annealing" [28]. On the other hand, the $(ATP)_7$ chamber participates stochastically in committing the chain to fold by protecting its folding nucleus. In this way, the GroEL $(ATP)_7$ state enables a productive initiation of the folding process that would be disrupted with non-negligible probability due to competing backbone hydration if the folding process were to take place in bulk solvent.

This part of the appendix shows that AI is capable of playing a transformative role in recreating and unraveling the molecular complexities of the cell and delineating their specific roles in assisting essential molecular processes that would probably not materialize otherwise or may do so in times that bear no relevance to life on Earth.

REFERENCES

1. Callaway E (2021) DeepMind's AI predicts structures for a vast trove of proteins. *Nature* 595: 635
2. Shaw DE, Maragakis P, Lindorff-Larsen K, Piana S, Dror RO, Eastwood MP, Bank JA, Jumper JM, Salmon JK, Shan Y, Wriggers W (2010) Atomic-level characterization of the structural dynamics of proteins. *Science* 330: 341–346

3. Brooks BR, Brooks III CL, Mackerell AD, Nilsson L, Petrella RJ et al. (2009) CHARMM: The biomolecular simulation program. *J Comp Chem* 30: 1545–1615.
4. Clark PL, Elcock AH (2016) Molecular chaperones: Providing a safe place to weather a midlife protein-folding crisis. *Nature Struct Molec Biol* 23: 621–623
5. Thommen M, Holtkamp W, Rodnina MV (2017) Co-translational protein folding: Progress and methods. *Curr Opin Struct Biol* 42: 83–89
6. Sorokina I, Mushegian A (2018) Modeling protein folding in vivo. *Biology Direct* 13: 13
7. Anfinsen CB (1973) Principles that govern the folding of protein chains. *Science* 181: 223–230
8. Jiang F, Wu YD (2014) Folding of fourteen small proteins with a residue-specific force field and replica-exchange molecular dynamics. *J Am Chem Soc* 136: 9536–9539
9. Fernández A (1999) Folding a protein by discretizing its backbone torsional dynamics. *Phys Rev E* 59: 5928–5934
10. Englander SW, Mayne L (2017) The case for defined protein folding pathways. *Proc Natl Acad Sci USA* 114: 8253–8258
11. Eaton WA, Wolynes PG (2017) Theory, simulations, and experiments show that proteins fold by multiple pathways. *Proc Natl Acad Sci USA* 114: E9759–E9760
12. Fernández A, Colubri A, Berry RS (2000) Topology to geometry in protein folding: Beta-lactoglobulin. *Proc Natl Acad Sci USA* 97: 14062–14066
13. Singhal N, Pande VS (2005) Error analysis and efficient sampling in Markovian state models for molecular dynamics. *J Chem Phys* 123: 204909
14. Zimmerman MI, Bowman GR (2015) Fast conformational searches by balancing exploration/exploitation trade-offs. *J Chem Theory Comput* 11: 5747–5757
15. Ekman M (2021) *Learning Deep Learning: Theory and Practice of Neural Networks, Computer Vision, Natural Language Processing, and Transformers Using TensorFlow.* Addison-Wesley, Boston
16. Fernández A (2021) *Artificial Intelligence Platform for Molecular Targeted Therapy: A Translational Science Approach.* World Scientific Publishing Co., Singapore
17. Fernández A (2020) Deep learning unravels a dynamic hierarchy while empowering molecular dynamics simulations. *Ann Phys* 532: 1900526.
18. Fernández A (2020) Artificial intelligence teaches drugs to target proteins by tackling the induced folding problem. *Mol Pharm (ACS)* 17: 2761–2767.
19. Zeiler MD (2012) ADADELTA, an adaptive learning rate method. *arXiv 2012, 1212.5701.* https://arxiv.org/abs/1212.5701
20. Laskowski R, Furnham N, Thornton JM (2013) The Ramachandran plot and protein structure validation. In *Biomolecular Forms and Functions: A Celebration of 50 Years of the Ramachandran Map* (Bansal, M. & Srinivasan, N., eds.), pp. 62–75, World Scientific Publishing Co., Singapore.
21. Thom R (1954) Quelques propriétés globales des variétés différentiables. *Comment Math Helvet* 28: 17–86
22. Weisstein EW (2021) Quotient space. From MathWorld: A wolfram web resource. https://mathworld.wolfram.com/QuotientSpace.html
23. Nemytskii VV, Stepanov VV (2016) *Qualitative Theory of Differential Equations.* Princeton University Press, Princeton, NJ
24. Lippert RA, Predescu C, Ierardi DJ, Mackenzie KM, Eastwood MP, Dror RO, Shaw DE (2013) Accurate and efficient integration for molecular dynamics simulations at constant temperature and pressure. *J Chem Phys* 139: 164106.
25. Nosé S (1984) A unified formulation of the constant temperature molecular-dynamics methods. *J Chem Phys* 81: 511–519
26. Hoover WG (1985) Canonical dynamics: Equilibrium phase-space distributions. *Phys Rev A* 31: 1695–1697

27. Martyna GJ, Tobias DJ, Klein ML (1994) Constant pressure molecular dynamics algorithms. *J Chem Phys* 101: 4177–4189
28. Thirumalai D, Lorimer GH, Hyeon C (2020) Iterative annealing mechanism explains the functions of the GroEL and RNA chaperones. *Protein Sci* 29: 360–377
29. Fernández A (2020) Artificial intelligence steering molecular therapy in the absence of information on target structure and regulation. *J Chem Inf Model (ACS)* 60: 460–466
30. Martin O (2014) Wrappy: A dehydron calculator plugin for PyMOL. *MIT License*. http://www.pymolwiki.org/index.php/dehydron

A.2 WAVE FUNCTIONS FOR UR-PARTICLES THAT DECODE ELEMENTARY PARTICLES IN THE STANDARD MODEL

The protocol for obtaining wave functions $\Psi : W \to \mathbb{C}$ for ur-particles decoding elementary particles is presented in Figures A.7 a,b.

(a)

$\{e_\mu, e_5\}_{\mu=1,\dots,4}$ *local basis vectors*

$$\vartheta_\mu = e_{\mu\perp} = e_\mu - e_{\mu\parallel} =$$

$$e_\mu - \frac{e_\mu \cdot e_5}{\|e_5\|^2} e_5 = e_\mu - \frac{g_{\mu 5}}{g_{55}} e_5$$

$$\Psi = \exp\left[\frac{iS(x,y)}{\hbar}\right] = \exp\left[\frac{ip_5 y}{\hbar}\right]\Phi(x)$$

$$\left[\left(\partial_\mu - \frac{i}{\hbar}mcg_{5\mu}\right)^2 + \left(\frac{mc}{\hbar}\right)^2\right]\Phi(x) =$$

$$\left[\left(\partial_\mu - ig_{5\mu}/r_0\right)^2 + r_0^{-2}\right]\Phi(x) = 0$$

4D Space-time

r_0

λ

α

$n\lambda = 2\pi r_0$

$e_{\mu\perp}$ e_μ

α $e_{\mu\parallel}$ e_5

(b) $g_{5\mu} = cos\alpha_\mu = cos\alpha$

Local metric components depend on type of particle at location, as the geometric dilution parameter α is associated with the particle with

momentum $p_5 = \dfrac{\hbar cos\alpha}{r_0}$ and wf $\Psi = e^{i\frac{cos\alpha}{r_0}y}\Phi(x),$

$$\left[\left(\partial_\mu - i\frac{cos\alpha}{r_0}\right)^2 + r_0^{-2}\right]\Phi(x) = 0$$

Light particle

Massive particle

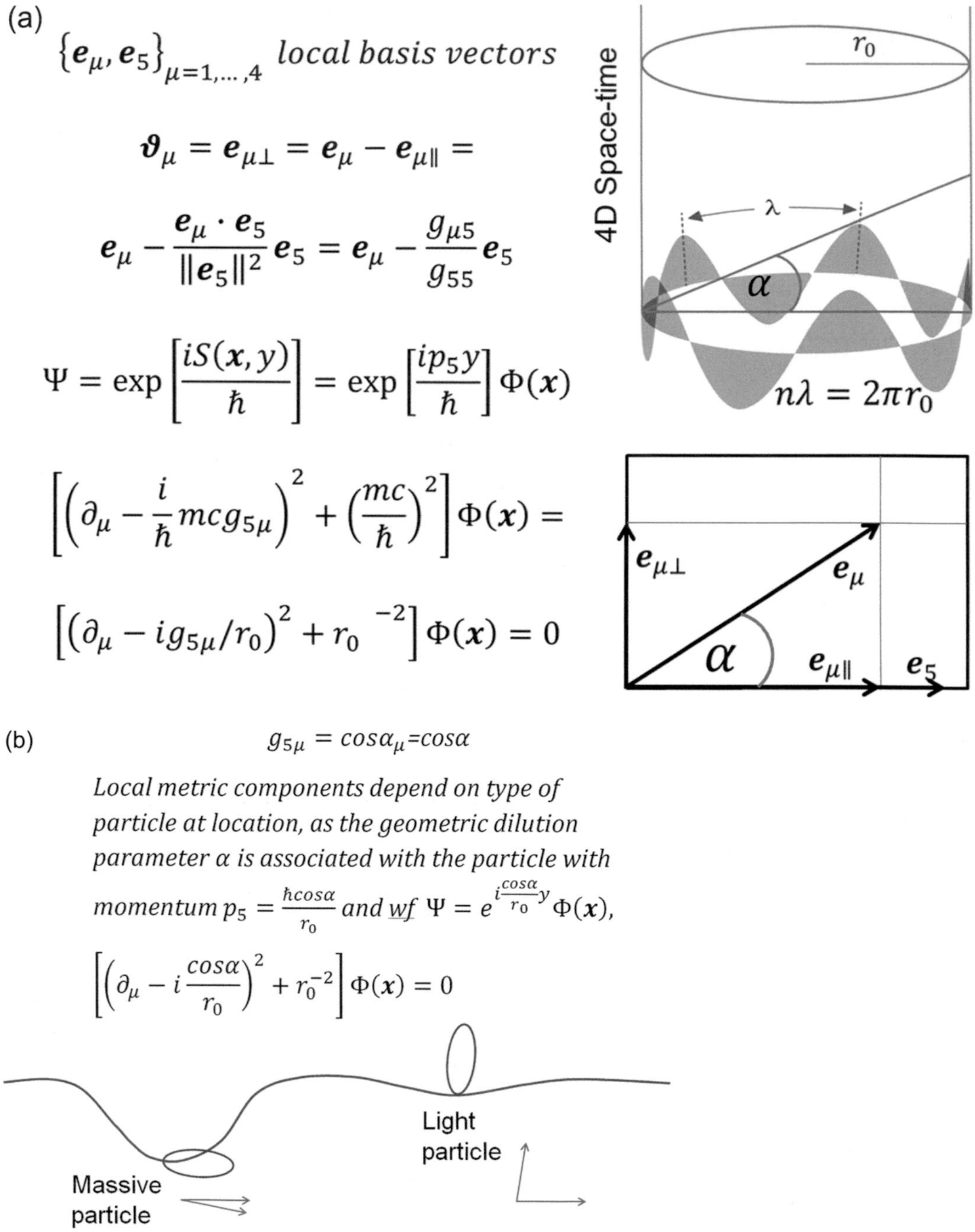

FIGURE A.7 (a, b) Protocol to generate the wave functions for ur-particles decoding elementary particles in the SM. Notation follows the main text in Chapters 4 and 6.

Index